Architecture Follows Nature
BIOMIMETIC PRINCIPLES FOR INNOVATIVE DESIGN

CRC Press Series in Biomimetics

Series Editor: Yoseph Bar-Cohen
*Jet Propulsion Laboratory, California
Institute of Technology*

PUBLISHED TITLES:

**Architecture Follows Nature—Biomimetic Principles
for Innovative Design**
Ilaria Mazzoleni

Biomimetics: Nature-Based Innovation
Yoseph Bar-Cohen

FORTHCOMING TITLES:

**Biomimetics and Ocean Organisms:
An Engineering Design Perspective**
Iain A. Anderson and Julian Vincent

Mechanical Circulatory Support for Heart Failure
Pramod Bonde and Robert L. Kormos

BIOMIMETICS SERIES

Architecture Follows Nature
BIOMIMETIC PRINCIPLES FOR INNOVATIVE DESIGN

Ilaria Mazzoleni
in collaboration with Shauna Price

CRC Press
Taylor & Francis Group
Boca Raton   London   New York

CRC Press is an imprint of the
Taylor & Francis Group, an **informa** business

Cover design by im studio mi/la, Richard Molina.

CRC Press
Taylor & Francis Group
6000 Broken Sound Parkway NW, Suite 300
Boca Raton, FL 33487-2742

First issued in paperback 2017

Version Date: 20121218

ISBN 13: 978-1-138-07669-3 (pbk)
ISBN 13: 978-1-4665-0607-7 (hbk)

Visit the Taylor & Francis Web site at
http://www.taylorandfrancis.com

and the CRC Press Web site at
http://www.crcpress.com

To my parents
& to our green valley

Contents

Foreword

All around us, we see the ingenious adaptive capabilities in the form and function of natural systems and the built environment. From single cells to multicellular organisms to ecological communities, we see architectural parallels in the function and form of the bivouac tent to the Louvre pyramid to Dubai's cityscape. We use the words exquisite and elegant without reserve to describe the wonders of nature, but can we always apply these superlatives to our tenements and washed out roads caused by how humans seem to intrude instead of blend with nature? One solution that raises our consciousness of the role man plays in the natural ecosystem of planet Earth is biomimicry. Copying can often be regarded as high praise and searching for life's analogies gives hope that perhaps humans may find where they fit by emulating natural interactions and balances, integrating complex functions seamlessly beneath a semi-permeable skin. The possibilities are endless but the way ahead is not spelled out. Biologists and non-biologists alike are becoming entranced by the potential of biologically inspired design. Yet those who have attempted to teach it note that there are few guidelines that make this process transparent. For our next steps, it has become clear that a deep knowledge of biology may serve us better than a superficial skimming that reduces the complex elegance to mistaken functions. Scientific knowledge is not that inaccessible anymore, given the new collections such as Planet Earth that show us some of the most dazzling feats of nature. Many of the mechanisms of these charismatic organisms are known and it is with great fun and fascination that today's practitioners of biologically inspired design are diving gleefully into a treasure chest full of natural solutions waiting to be realized in human design.

To tempt us even further is this treatise by Ilaria Mazzoleni. Be amazed by the known biomimetic designs that exist, the known natural systems left to be translated, the exponential rate of technological advances that will enable you, too, to be a bio inspired designer! The information is at your fingertips but the translation will be made easier by making a friend with a biologist. Go ahead. Let's match the complexity of nature with a complexity of expertise in an interdisciplinary team needed for bio inspired design. We have the knowledge; we just need more practice using it. Take a closer look at the life around you and one day you'll see much more of it mirrored in the tools we use, the houses we live in, the rules we rely on. This new intuition gained by using this approach may actually help us find harmony with nature.

Jeannette Yen
Professor, School of Biology
Director, Center for Biologically Inspired Design
Georgia Institute of Technology, Atlanta, Georgia

Acknowledgments

I am deeply connected to two places in this world: Los Angeles, California — with its expansive horizon, deep ocean, blue sky, desert light, and sunny days; and Val Taleggio, Italy – with its immersive green landscape, deep mountainous canyons, white peaks, and fresh air. These natural environments have shaped my profound love for nature, but most importantly have connected me to the influential people who have molded me into the person I am today. My deepest appreciation goes to them for supporting the genesis and development of this book and celebrating its completion. Without them nothing would have been meaningful to begin with, and because of them I hope to modestly contribute in shaping a better world.

First and foremost, I am truly grateful to Shauna Price; without her contribution the book would not have been possible. Our collaboration started with her participation in my biomimetic courses, where she instilled fundamental concepts which have been refined further in the writing of this book. She kindly nurtured my interest in learning about the governing principles of nature and welcomed the idea of a collaboration between biology and architecture. Fantastic and nourishing ideas have flourished with help from our team at im studio mi/la: Richard Molina's dedication has been instrumental with the management and graphic integrity of the book's complex and rich material. A special thank you to Juan Miguel San Pedro for his positive energy in editing all diagrams and drawings of the twelve projects included in the book.

I would like to express my gratitude to Janine Benyus and Steven Vogel for their positive feedback on preliminary drafts of the book which encouraged me to show it to Yoseph Bar-Coehn. Yoseph's immediate enthusiasm in wanting to collaborate as the editor and proposing the book to CRC Press has been a very gratifying experience. I am thankful to Jeannette Yen for her kind words in the foreword.

Since 2005, the Southern California Institute of Architecture (SCI-ARC) and the institute's directors, chairs and coordinators have all graciously supported the notions and implementation of a variety of biomimetic seminars I have proposed and taught. Over these years various colleagues and interdisciplinary collaborators have participated in the seminars through a series of lectures and critiques which have created a nurturing, collaborative atmosphere which we all benefited and grew from. Thank you to the following people: Berenika Boberska, Juan C. Portuese, Yo Yoshima, William Howard, Lorraine Lin, Steve Ratchye, Prof. Geoff Spedding, Borja Mila, Jaime Chavez, and a number of other biologists who specifically contributed to the editing of Part II.

Different segments of the manuscript have found the keen eyes of readers and editors such as architect Sarah Graham and environmental lawyer Suzie Lieberman. The patient guidance and beneficial input provided by Sarah Dennison has been instrumental through the process; being both a biologist and architect, no one better could interpret my thoughts and guide me through the maze (and my obfuscation) of the English language.

Useful ideas and clarifications emerged from conversations with Alessandro Mendini, Stefano Boeri, Pam Thompson, Mohamed Sharif, Walid Soussou, Deborah Gloria, Diego Terna, Asli Suner, Jan Ipach, Claudio De Fraja, Massimo Garbuio, Fabrizio Gallanti, Jeffrey Landreth, Anne McKnight, and Arun Krishnan. Throughout the process they graciously supported me, encouraged me, and aided with brainstorming sessions.

The vibrancy of the book lies not only in its text, but also in the carefully selected imagery. The team early on decided to find images of animals photographed by scientists. The long and intense search for these images became a journey all on its own, allowing us to become acquainted through email with great people dedicating their life to field work. It has been an adventure to reach out, learn what they are doing, and encounter such generosity in sharing their photographs. I truly loved this process – virtually traveling with them, immersing myself in their observations, and validating that great discoveries are still ongoing in the twenty-first century. Thank you all!

I would like to conclude by thanking my family and friends, all of whom have supported my desire to write this book for years and continuously encouraged me forward.

I wish the reader a pleasant time of discovery!

Project Credits

I gratefully offer thanks to the students who enthusiastically contributed to the book, in particular those who put extra effort into their original seminar project. You know who you are!

Most of the projects were undertaken by SCI-Arc students and initiated under various designations of the seminar "Biomimicry: Innovation in Architecture Inspired by Nature" throughout the academic years of 2010–2011. An important and fundamental role is attributed to the interdisciplinary editorial team, who have read, analyzed and edited the projects – transcending their strengths and coherency. To each and all of them, my profound appreciation, in particular biologists Graham Slater, Ryan Ellington, Ryan Harrigan, and architect Stacy Nakano. Thank you.

Project Team Members

Urania moth – **Benedetta Frati, Nir Zarfaty**
Violet-tailed sylph – **Joanna-Maria Helinurm, Alina Amiri**
Lettuce sea slug – **Ana Munoz, Ryan Hopkins**
Side-blotched lizard – **Juan Miguel San Pedro, Alex Nahmgoong, Yuan Yuan**
Snow leopard – **Joakim Hoen, Mamoune Ghaiti**
Polar bear – **im studio mi/la, Ilaria Mazzoleni, Alessandro Colli, Richard Molina**
Banana slug – **Astri A. Bang, Maya Alam, Janni S. Pedersen**
Dyeing dart frog – **Erin Lani, Jordan Su**
Ochre sea star – **Paul Mecomber, Adrian Ariosa**
Namib desert beetles – **Emily Chen, Carlos Rodriguez**
Tree pangolin – **Ross Ferrari, Thomas Carpentier**
Hippopotamus – **Sarah Månsson, Worrawalan Raksaphon**

Preface

Biology is a rich discipline. Its influence spreads far beyond the field of science and extends into seemingly unrelated fields like architecture. The term "biomimetics" has been in use since the 1960s when Otto H. Schmitt defined it as "biology + technology" but applied it mainly within the field of engineering. In the field of architecture, however, biomimicry has only been used since the early 2000s, reconsidering biomimetics as applied to design.[1] While biomimicry has been typically limited to imitating the morphological aspects of the biological world, its potential to reveal the functional aspects has been largely overlooked. This book, by providing an understanding of fundamental biological concepts of evolution and, in particular, adaptation, proposes a new way of looking at the functional aspects of nature as sources of inspiration for improvements in application to building design and form.

In this book the biomimetic approach to design is seen as an opportunity to bring attention to the unique capacity of nature to work in a systemic way. In fact, everything in nature is interconnected to the degree that some scientists consider Earth a single ecosystem or biosphere discouraging discrete views of smaller scale systems that exhibit discontinuity and separation. Nature's systems are dynamic, in flux, and in constant transformation, subject to the laws of physics. Interconnectivity also inherently involves the concepts of coexistence and richness in biodiversity. Natural systems depend upon the diversity of their elements, which ultimately create balance within an organism or ecosystem. In architecture we are accustomed to thinking of elements in isolation. And while this allows concentration on details, it misses the possibility of working within a model of ecological, dynamic contextualization. The study of interconnectivity of individual elements in nature can lead to a significant shift in contemporary architectural thinking. If inspiration is taken from biology it is easier to understand nature's importance and to find meaningful design examples. In design it is always useful to begin with a simple, holistic building concept. This book is concerned with building envelopes and their multifunctional connective potential. Building envelopes may be thought about in terms of both the interior realms of the building and their capacity to address the forces of the larger exterior environment. From part to whole, from local to global, from micro to macro, everything is connected and connectable, in constant, harmonious flux and unfixed in time or space. From buildings to communities we can design complex ecosystems rich in interconnected solutions which offer an infinite range of possibilities. Ecosystems can generate concepts for allowing all elements to coexist with each other. Learning from nature's efficient interconnectedness allows designers to consider opportunities for multifunctional uses and streamlined design solutions. This can ultimately reduce and rebalance our footprint on Earth while producing an integrated built and natural ecosystem, eliminating the separation between artificial and natural. Moving from designing buildings in isolation to considering them within larger networks involves a simultaneous understating of a multitude of factors. Part of this transformative shift in thinking also involves reimagining the role of architecture in terms of its restorative potential. Restoration of natural systems could become a significant contribution of architecture.

The case studies presented in the book are explorations of the animal kingdom that consider the climatic and ecological contexts in which the selected animals evolved. Adaptations, which allow animals to survive in their habitats, provide lessons that, in turn, are translated into designs for the built environment. Such investigations focus on the analysis of various animal skins and skin coverings. Animal skins are one of the major systems for which architecture can draw inspiration from biology. There are, however, many others, such as plant species, that might be considered, and the methodology proposed in this book can be used to guide one through the process in an analogous fashion. Skin is a complex and remarkably sophisticated and variable organ, providing animals with protection, sensation, and heat and water regulation.

The term "skin" is used in this book on a general level to refer to any animal covering, including fur, feathers, scales, exoskeletons, and shells. Skin is understood as an interface, transcending its surface, giving the appearance of something that separates, but instead acts as a threshold or boundary, allowing for interaction with the elements in multiple directions, scales, and timeframes. Similarly, building enclosures provide building inhabitants an interface with environmental elements such as weather, noise, and sunlight. Architectural enclosures furnish great opportunities to take into consideration dynamic local environmental conditions, creating the potential to use the conditions as resources to be enhanced and supported rather than simply as elements to conceal or overcome. Biomimicry is a rich tool for designers to innovatively integrate the local environment into their projects, supporting a more sustainable way of building and living.

In its use to date, biomimicry has been predominantly applied to form in architecture. Instead, this book is interested in focusing on biomimetic applications to architecture in both form and function. The methodology is deployed through a series of sample projects that resulted from a collective investigation that took place during the course of a seminar born from ongoing professional research by the author and her academic collaboration with students at the Southern California Institute of Architecture and biologists from the Department of Ecology and Evolutionary Biology, University of California, Los Angeles.

The book is organized into two parts. Part I introduces and describes the principles of design that take inspiration from nature as well as the fundamental biological concepts that can inform architecture. The biomimetic methodology, developed by the author, is introduced in Part II which explores four sets of case studies, each of which investigates a particular function of skin: communication, thermoregulation, water balance, and protection. The integration of these functions with other internal body systems is also explained. The chapter on each animal has two components. The first introduces and analyzes a selected animal and its skin functions. The second assimilates biomimetic applications within a theoretical "proto-architectural" project, where the prefix *proto-* indicates that the projects here presented are at their conceptual stage of development.[2] Proto-architectural thus defines a key moment of design, open to discourse, experimentation, and environmental innovation. The examples are intended as a springboard for applying the functional analysis of an animal's biology to the design process.

The new methodology provides a path for drawing design inspiration from nature. It considers architecture beyond the aesthetic or functional, and begins to explore the conceptually strategic. The book proposes a novel way of looking at the environment and provides cues to undiscovered inspirations for a variety of audiences: for designers and architects, it provides biological inspiration for further flexible and dynamic design, and templates for them to derive other species examples, and acts as a springboard for exploring broader applications beyond envelopes to applying whole ecosystems to entire buildings and communities. For engineers, it is hoped that the book might inspire new building technologies and design of materials. The book can be used by biologists to continue researching nature and to make their findings available and understandable to designers and architects. For environmentalists, it may be used to help overturn the stereotype that ongoing building construction must inherently be ecologically damaging, neither capable of maintaining nor restoring land. For others, the book looks to promote greater appreciation of biodiversity and incentives for conservation by bringing attention to the vast variety of species and their adaptation to their environment. *Architecture Follows Nature—Biomimetic Principles for Innovative Design* seeks to inspire a shift in thinking about the built and unbuilt worlds by illustrating that they can be intelligently integrated, and that bio-inspired design might not only help the environment but also be beautiful!

PART I

1 Theoretical Framework

"Nature does nothing uselessly."

- Aristotle

Introduction

What is nature? What is our relationship with nature? How can we change our relationship with nature to mitigate the impacts we have brought upon its systems and ourselves?

The word "nature" has its roots in the Latin word *natura*, which in the classical era meant "birth" or "begetting." "Nature" is also conceptually close to the Greek word *physis*, which refers to a material system that exists and grows. Accordingly, nature is by no means static and is constantly regenerating and transforming itself. The interconnectedness of natural systems on Earth supports life, which evolves at myriad speeds and scales so far not found anywhere else in the universe. Interconnectedness is the theoretical concept for this book, defined as the integration of nature's solutions with innovative problem solving for man-made environments. The Earth's natural resources are finite and the planet is increasingly vulnerable to human activities, making apparent the responsibility we have to lessen our impact on it.

The process of evolution and the resulting adaptations have allowed life to sustain itself for millenia. But the increased pace and scale of human activities has unknown consequences for the balance of systems that allow all species, including our own, to thrive. Sustainable design is a way for us to begin to harmonize man-made structures with the natural environment. Biomimicry can help us change our perception by looking to nature as a source of functional and aesthetic solutions rather than as a source of obstacles to overcome.

Technological developments have generally been born from the mindset that nature can be harnessed to suit human needs. Historically, humans have used ingenuity to develop tools that have allowed us to transcend the environmental stresses threatening our survival. From the spear to the wheel to the building, our inventions have allowed us to catch prey, travel further distances, and create shelter in otherwise inhospitable places. Architectural design has grown out of the need for shelter and expression, and today we have the ability to design and engineer buildings with technologies and strategies that provide high levels of comfort in any climate. Although technology has been very successful in facilitating our survival, technology has also resulted in the unintended consequences of resource depletion, pollution, and climate change. During the past few decades it has become clear that a new way of thinking is inevitable to address these issues.

The application of the life sciences in building design has so far been limited largely to the imitation of organic form. Biologists and ecologists have studied the environmental influences on animal and plant physiology and behavior, but translating these observations and analyses into architecture has been largely unexplored. The resilience of species in a particular habitat can provide valuable lessons for long-lasting design. Just as animals have systems, such as skeletal, circulatory, immune, digestive, communication, and sensory, so too do buildings have systems of structure, circulation, protection, energy and water use, communication, and thermal regulation. Viewed as a network of internal systems interacting with its surrounding environment, which is in turn part of a larger global network of systems, the building can find inspiration from an animal's interactions with its ecological realm.

This book shows how explorations of the animal kingdom can help us consider the climatic and ecological contexts in which the animals selected as case studies have evolved. Building envelopes share much in common with animal skins and can borrow an enormous amount of information from them. Like an animal skin, a building enclosure acts as an interface, allowing for interaction with the elements. Enclosures have the potential to act like natural filters with the environment, rather than barriers, by being reactive to the direction of local winds, solar orientation, and humidity. One can imagine a building that reacts to changes in weather by altering its shading configuration or water-capturing abilities. Architectural enclosures, once reimagined as flexible and reactive, can furnish opportunities for taking into consideration dynamic local environmental conditions that might allow buildings to co-exist with nature, preventing nature's degradation and perhaps contributing to its restoration.

How can architects and designers move beyond the formal imitation of nature to more sophisticated, nature-inspired, performance-based building design? Successful nature-inspired design would need to include teams of collaborators from multiple disciplines, not only engineers, contractors and owners/users. Physicists, biologists, ecologists, and other scientists would also be included as part of the team for their expertise in understanding natural solutions. Biomimicry stresses the interconnectedness of systems to solve complex problems; similarly, the integration of varied disciplines yields fertile ground for comprehensive designs to address the array of environmental issues in which our buildings are constructed and operated. Smart solutions derived from examining nature have the potential to harmonize with the environment, rather than exploit it.

Architects, as artists, constantly look to new and emerging ways of exploring and developing ideas to create buildings that not only function well, but also express the culture and technologies of their time and set the standard for future ways of living. Contemporary Parametric Design, for instance, is an area of study where designers manipulate the algorithmically coded parameters of digital models to organically generate complex geometries, which are then used for building facades and structures. Furthermore, nanotechnology has provided the potential to integrate carbon nanotubes in building materials to create surfaces that react to our touch. Through experimentation, imagination and creativity, biomimetic design is a new direction in which to discover and transform the way we, and our built world, relate to the natural world. There is a growing movement to RE- : REthink, REduce, REpair, REuse, REcycle and REimagine the ways in which we

Human shelters carved out of the rock in Cappadocia, Turkey, 4th Century. (Photograph courtesy of A. Suner)

inhabit our planet. It is the motivation for the studies described in this book.

How Nature Inspires Architecture

Architecture has always inserted itself into and interacted with the natural environment. Essentially, architecture provides shelter in nature to protect its inhabitants from nature.

Historically, form has been the primary source of inspiration from nature, ranging from simple formal influences to the more symbolic translation into architectonic language. The Ancient Greeks fashioned the ornamentation of their columns and temples on local plant life to symbolize nature. Today designers are digitally developing an architectural vocabulary (most of which has yet to be constructed) which resembles forms found in nature. Can we find contemporary ways to "Touch This Earth Lightly"[3] through biomimetic design and the analysis of an animal's physiological and behavioral adaptations to local environmental conditions? Our knowledge of nature should be used more fundamentally as a resource for sustainable design and for the realization of more efficient buildings.

Nature offers additional insights for architecture in its ability to act locally but with an indirect ability to have a global influence. This is illustrated by the interconnectedness of ecosystems. Ecological landscapes have no fixed boundaries, rather they have edges that are influenced by surrounding communities. This creates a sequence of locally refined solutions and a network of connections that allow for adaptation to occur over time within each element of the community or ecosystem.

From buildings to neighborhoods we can design complex ecosystem-like entities, rich with interconnected solutions, which offer an infinite range of possible results. Ecosystems are rich in opportunities for allowing each element to coexist with the others. Learning from nature's efficient interconnectedness allows designers to consider the potentials for multifunctional responses. It also offers some examples, as seen in animals, of streamlined design solutions which might serve as models to ultimately reduce and rebalance our footprint on Earth. By embracing bio-inspired design processes, opportunities arise to help us develop man-made environments in harmony with nature, beginning to eliminate the separation between the built and the living realms.

Historical Interplay of Bio-Inspired Architecture among Science, Art and Design   Archeological sites testify to the fact that nature has been a source of observation and impetus for new design since ancient human civilization. Examples of Etruscan artifacts and ancient drawings found in caves demonstrate the intense need for societies to worship nature. Followers of Greek polytheistic religions worshipped a constellation of gods and goddesses named after the natural elements. The important Greek philosopher and polymath Aristotle (384–322 B.C.) put nature at the center of his scientific studies. In his *Historia Animalium* he describes many zoological phenomena. In the ancients' architecture, nature's forms are used symbolically and metaphysically. Nature has always been something to observe, represent, respect, and worship. However, with the major exception of Leonardo da Vinci, it was not until the 19th century that thinkers made the leap from "mere" observation to application. Leonardo could be considered the first biomimetic designer.

With the Age of Exploration, and increasingly after the discovery of the Americas (1492), an influx of European naturalists documented their field observations in the form of drawings replicating nature. As a result, science progressed using tools borrowed from the art world. History's greatest naturalists, such as Leonardo da Vinci (1452–1519), Konrad Gessner (1516-1565), and Ulisse Aldrovandi (1522–1605), among others, produced stunning informative drawings. Gessner's *Historiae animalium*, published in Switzerland around 1555, is considered the first encyclopedic work dedicated to documenting all known animals, particularly through the inclusion of illustrations, drawn mainly by Lucas Schan. The book generated such an impact that a few years later all of its illustrations were collected in a separate book, *Icones Animalium* (1560). We can attribute the importance first given to direct personal scientific observations to Aldrovandi. His watercolor and tempera illustrations, together with a copious collection of drawings prepared in his studio by artists hired and supervised by him, depicted real, directly observed organisms, accurately studied in all their external as well as internal details.

Owl illustration. Ulisse Aldrovandi, *Owl, Ornithologiae*, engraving, 1599. (Image courtesy of Wikimedia Commons)

Bat species illustrations. Ernst von Haeckel, Chiroptera, *Kunstformen der Natur*, plate 67, 1904. (Image courtesy of Wikimedia Commons)

In the following century, during the Age of Enlightenment, 1650–1800, scientific explorers and voyagers such as Alexander von Humbolt (1769–1859) explored the coasts and the inlands of Central and South America, documenting his discoveries with beautifully detailed drawings. von Humbolt had a great influence on the explorative work of Charles Darwin, while Ernst von Haeckel, (1834–1919), biologist and artist, beautifully helped document Darwin's observations graphically. The intellectual world became fascinated by the pioneering and explorative findings, so much so that Denis Diderot (1713–1784), in his *Encyclopédie*, dedicated a good 12 volumes to "planches" (plates).

Less tangible and seemingly invisible aspects of nature also inspired the development of technologies that allowed further investigation. The invention of the microscope in the late 16th century in the Netherlands and later the compound microscope by Galileo Galilei (1564–1642) in 1625 allowed scientists to study the incredibly close and small as well as the distant using the same technology. Today, with the aid of the electron microscope, we can observe the fine structure of a single cell, and with nuclear magnetic resonance (NMR) spectroscopy, observe protein structures. Robert Hooke (1635–1703) is considered the first to bring to the public astonishing microscopic images from the invisible to bare eye from the world of nature.

These early naturalists and thinkers all demonstrated that through analysis and scientific research, it was possible to create graphical renditions of the natural world. It was only Leonardo da Vinci who, contrary to his contemporaries, developed the observations into design ideas and concepts. Several of da Vinci's famous drawings demonstrate the shift from mere inquiries to a world of human creation and design. In fact, Leonardo may be regarded as the first biomimetic designer, considering his investigations on the flight of birds, which developed into the invention of the first flying machine – the Ornithopter. Another critical da Vinci contribution is the elaboration of Vitruvio's *de Arquitectura* text, from which he extrapolated and developed the geometric relationships of the human body to the pure geometric forms of the square and circle. This advanced way of "observing" opened the world to what we call the "relational aspects." Leonardo's drawings provided an essential link to early relational observations in demonstrating that everything in nature is interconnected, and

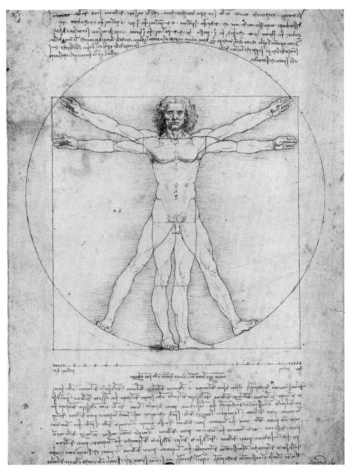

Leonardo da Vinci, *Vitruvian Man*, pen and ink on paper, c. 1487. (Image courtesy of Wikimedia Commons)

that clear, relational rules from nature can be applied through geometry. In this way, creativity comes to play an important role in the mediation of scientific reason, function, and formal expression through drawing and design.

Concurrent with the great scientific discoveries, art flourished during the Renaissance. Interestingly, painters also started to feed their imaginations with naturalists' observations of landscapes and nature. The art world developed a deep fascination with the natural world. Artists' clients, who were often removed from bucolic landscapes, requested their artists bring those idyllic

Giuseppe Arcimboldo, *Water,* oil on limewood, 1563–4. (Image courtesy of Wikimedia Commons)

views into their homes. While some provided realistic visions of the natural landscapes, others applied a freer level of interpretation and creativity, distorting and recomposing their visions into a multiplicity of fantastical variations. Art started to position itself between the recording of present realities and the imagining of other places and worlds, thus enhancing spectators' fantasies and visions. The creative genius of such inventiveness can be found in the exceptional work of Giuseppe Arcimboldo (1527–1593). His fantastic human-like portraits comprised of assemblages of animals or plant species are exemplary of the inspiration science and nature provided to art, and how the imagination of the artistic mind could recompose such elements in creative and surprising ways.

The 19th century saw a proliferation of engineering and other applied sciences, with bright minds applying their efforts to the testing and discovery of many devices. A big shift happened in 1857, when Jean-Marie Le Bris, during one of his long sailing trips following the flight of an albatross, designed and built the first bio-inspired flying machine, the "Artificial Albatross." Another milestone came with the age of modern robotics in the 1950s. This time defined a critical threshold of a new branch of engineering in which scientists were developing bio-inspired devices. Over time robots have become more and more similar to living organisms.

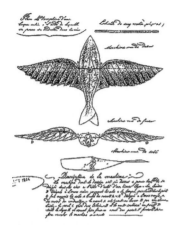

In 20th century architecture there are few relevant case studies designed by architects informed by nature. Influenced by his mentor, Luis Sullivan (1856–1924), Frank Lloyd Wright (1867–1959) put at the center of his career his interest in nature. He considered himself the instigator of organic architecture. In his book *An Organic Architecture* (1939), he describes how he believed not only that every building should grow naturally from within its surroundings, but also how the building's design should be carefully thought of as if it were a unified organism, in which each element of it relates to the other, in which each element relates to the other, similar to ecosystems in nature. Villa Mairea (1939), designed by the Finnish Alvar Aalto (1898–1976), is one of the finest examples of nature's influence in architecture. The forest that surrounds the house becomes the driving element for the conception of the interiors made of irregular columns and posts. The interiors resemble the diversity and beauty of nature, from materials to forms, in the intent to dissolve the separation between the indoor and the outdoor environments, as in Alto's words "nature is the symbol of freedom."

(Top) Illustration of Artificial Albatross. Jean-Marie Le Bris, *Brevet d'invention,* 1857. (Image courtesy of Wikimedia Commons)
(Bottom) The living area immersed in a forest of columns. Alvar Aalto, Villa Mairea, 1939. (Photograph courtesy of A. Carr)

Buckminster Fuller (1895–1983) paid close attention to nature and its governing systems, as he understood that humans exist in connection with the rest of the living world. Jean Prouve's (1901–1984) Maison Tropical (1949) and the later Maison du Sahara (1958) not only explored ideas of prefabrication and lightness, but are especially important for their design for extreme climates. Frei Otto (b. 1925), in his prolific career, concentrates on finding the basic principles of structures in nature. His contribution and influence are still, today, predominant in the fields of minimal surfaces and complex geometry. In the late 1950s the Metabolists, a group of

The Metabolists' most significant built contribution that synthesizes notions of exchangeability and growth. Kisho Kurokawa, Nakagin Capsule Tower, 1972. (Photograph courtesy of M. Nesbit)

Japanese architects led by Kenzo Tange (1913–2005), began looking at cities and buildings as expandable, flexible structures, capable of organic growth and transformation over time. Evidence of the Metabolists' notions are significantly demonstrated in Kisho Kurokawa's (1934–2007) infamous Nakagin Capsule Tower, which embodies an explorative and adaptive design attitude for a building as organism. After decades of postmodernism, attention has been refocused on nature's morphologies. Responsibility for this trend can be attributed to, among others, the work of D'Arcy Wentworth Thompson (1860–1948). The rediscovery of the 1917 book *On Growth and Form* by the mathematician has influenced generations of architects and designers including organic morpho-architects. His meticulous work looked at the correlations between biological forms and mechanical phenomena, and his descriptions of the interrelation between form and growth helped biologists, architects, and engineers find fruitful starting grounds for collaborative explorations.

The attention brought to bio-inspired design in the last 15 years has allowed the emergence of a new group of designers and architects interested in interpreting, participating in, and collaborating in the development of architecture which is environmentally and ecologically sound, as well as atmospheric, sensible, and smart. Today there are many contributors to the advancement of bio-inspired design and speculative projects within and outside the academic realm. Only a few examples can be truly considered "biomimetics," with even fewer built examples in architecture. Increased attention to the subject matter has reached a critical mass for further development. Because we are facing some of the most unprecedented environmental challenges of our time, we need architectural responses that are similarly unprecedented, flexible, adaptive, and performative in function. Biomimetics, by finding direct inspiration from nature, provides unmatchable lessons to designers and architects.

"A building should be designed so as to minimize the use of new resources and, at the end of its life, to form the resources for other architecture."

- Robert & Brenda Vale

Contemporary Challenges and Interests Climate change is arguably the most far-reaching, unprecedented environmental challenge of our time, and has been recognized as such by the

United Nations, the national science academies of all major industrial nations, a majority of the world's governments, and the mainstream public. Although its long-term effects on the environment are not yet clear to us, it is apparent that the extreme shifts in climate wrought by humans have real repercussions on our ecological equilibrium. The media constantly remind us of the damaging effects that the perennially melting glaciers, rising sea levels, flooding, drought, fire, heat waves, catastrophic weather events, altered ecosystems and biodiversity loss, and land and air pollution have upon our health and economic and political stability.

We are also accelerating the extreme loss of biodiversity through the loss of habitats caused by deforestation, urbanization, water, land and air pollution, and other forms of natural resource depletion. The extinction of species creates large ecological unbalances. When one species disappears, other species will be affected, provoking a chain reaction of events. Organisms are facing environmental changes at a rate that is highly accelerated compared to the normal course of evolutionary change, and therefore, many species cannot adapt to the unprecedented ecological pressures that humans have caused. But what is currently induced is a mass extinction of species, thousands of which still have yet to be discovered and thousands more classified, jeopardizing entire habitats, the species as indicators of habitat health, and the services these ecosystems provide to humans. The International Union for Conservation of Nature (IUCN) estimates that we are losing species due to anthropogenic pressures at least 1,000 times the natural rate[4]; 19,265 species of the 59,507 assessed to date are threatened with extinction.[5] We urgently need to change the way we interface with the natural environment.

As the built environment also significantly contributes to climate change, it is urgent, and logical, that we consider an integrated design approach incorporating nature's time-proven lessons. The architect's and designer's goal is to develop ways to appropriate and reuse nature's resources responsibly.

Our challenge today is to balance or eliminate waste by developing ways to build with limited environmental impact and without excessive abuse to the natural world. This applies to how we design and build, and to how we occupy, maintain, and operate our buildings. Nature is very complex and highly dependent on closed-loop cycles. In this homeostatic state, one's waste is another's food source. Waste is one of the major causes of resource depletion and only truly happens when we interrupt natural cycles.

Bio-inspired design is a transition toward revitalizing, revamping and reinventing our existing environments. This in turn provides an opportunity for designers of both new and old construction to shift their focus from developing buildings in isolation to considering them as a component of the larger network of systems intertwined with the multitude of factors making up the environment as a whole. Performance-based and whole-system approaches offer infinite possibilities to creatively reinvent, reprogram, distort, manipulate, rearticulate, and reshape the environment that surrounds us. Through our vision of a more compact, more responsible, and more creative way of coexisting, we can come closer to living in a world of equilibrium. The understanding and design of the building envelope, under particular examination in this book, offers a great opportunity for addressing such issues.

Performance-based design is necessary in order to lessen our ecological footprint. Innovators are focusing on performative design techniques used to optimize life-cycle building performance in an integrated, holistic manner. Integrated systems design coordinates complex systems intended to maximize building functionality while providing for human comfort. The integration of advanced building systems (e.g., envelope, mechanical, electrical, lighting, and plumbing) combined with sustainable design practices is our best way forward in this regard. Performative design can encompass complex geometry, parametric and algorithmic design while moving the process beyond the mere formal observation of natural forms and patterns. Thus, the goal of this process is the development of meaningful, flexible and adaptable relationships between systems from which architecture and its processes can emerge.

Working holistically does not eliminate the need for step-by-step process-based design, a hierarchical approach to addressing major topics such as climate and design principles (orientation, program and space zoning), envelopes and passive strategies (performative and adaptive parametrically complex envelopes, thermodynamic principles, and innovative materials), active systems (building systems design), and generative systems (renewable energy sources). Today quantitative and qualitative performance-based analyses offer a comprehensive virtual approach to validating the process. Digital tools and assessment methods allow not only for preventive testing and methodology validation but also the generation of the building form.

Systems integration starts with the design of the envelope, one of the most important components of a building, as it incorporates most of the architectural and building engineering disciplines. The envelope connects and separates, and acts as a filter between the exterior world and the internally controlled environment. It mediates and enhances all relationships between natural elements and the conventional notion of human comfort. An advanced understanding of building and environmental design concepts, principles, and strategies is necessary in order to identify appropriate building systems for different climates and building occupancies. Extensive analysis, testing and systems optimization are required elements of a performative building.

"I have thought exactly the opposite. Jungles and grasslands are the logical destinations, and towns and farmland the labyrinths that people have imposed between them sometime in the past. I cherish the green enclaves accidentally left behind."
- E.O. Wilson

Nature and the Built Environment Presently, there are very few places in the world occupied "lightly" by man, such as places where tribal communities remain part of natural ecosystems. And there are even fewer places left completely untouched by humans.

One can establish a quantitative relationship between the built and unbuilt environments. However, such quantification alone does not necessarily lead to a definitive answer we can implement to produce real change. We must consider a qualitative assessment of this quantitative relationship if we are going to productively shift and reverse the currently large human ecological footprint. A constructive attitude toward conservation and restoration policies starts with the healing of sites' ecological and architectural systems at the master planning scale. We must modify the way we design our buildings so that they will not contribute to the

Scarred landscape and debris of dead fish off the shores of the Salton Sea, CA. (Photograph courtesy of R. Cha)

depletion of natural resources, but rather work to reintroduce resources back into ecosystems, ultimately benefitting the environment at large.

An example of a restorative approach to design can be found in the project "A Model Community at Salton Sea" by im studio mi/la and collaborators. The traditional model of growth is a zero sum game with discrete land use typologies, where growth in one area can only happen at the expense of another. This project provides a paradigm shift in accommodating growth – by capturing and integrating systems that are cyclical in nature and time, with each cycle rejuvenating and healing the surrounding ecology rather than eroding it. The approach is holistic, in that it considers interrelationships between all processes fundamental to sustaining life and preserving nature – water and energy cycles, agriculture and seasonality, production and the exchange economy, as well as the social needs of a multigenerational community. The applied strategy, however, is hinged on the notion of restoring scarred landscapes, making them givers of life, and enhancing their integration into the surrounding ecology.

As a result of untrammeled population growth, the development of cities and degradation of rural sites have confined nature to an increasingly smaller and limited area intended for conservation, monitoring, and biodiversity preservation – both outside of and within small and large cities. The concept of nature in urbanized areas has transformed and redefined our understanding of what nature is and how to interact with it. The question is if city-dwellers still maintain a relationship with nature in their urban lives. And if the sense of beauty we associate with nature is in contemporary living something we relate to only in a romantic, bucolic, and melancholic way, rather than in an interactive, daily, practical way.

Nature is dynamic and adaptive, thus allowing for restoration. Healing degraded areas is the first step toward a healthier relationship between nature and the urban environment. The second step lies in architecture's long- and short-term response to nature's ever-changing conditions; architecture needs a new context in terms of environmental factors that define a place, beyond the traditional contexts of physical surroundings, geometries, architectural styles, and stylistic traditions.

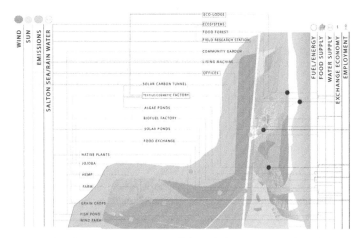

Master plan depicting inputs on the left, and the outputs produced by the community filtered through the system on the right. Model Community at Salton Sea, CA, 2006. (Image courtesy of im studio mi/la and collaborators)

The notion of aesthetics also plays a substantial role in incorporating nature into the built environment, as it establishes emotional attachments and garners respect. In human society, beautiful places tend to be valued more. By designing aesthetically pleasing things, we can help their conservation. Issues of aesthetics in architecture have been discussed since the time of the ancient Greeks, a society highly interested in such philosophical aspects of life. In science, beauty is also often present but discussed in different terms, such as through laws and mathematical formulae. The semiotician Umberto Eco, in his book *History of Beauty*, offers a peculiar reading of how this concept often relates to the feminine attributes of curves and grace. But architects and scientists seem to disagree on the importance of simplicity versus complexity, elegance, and symmetry as elements that would conclusively define what is pleasing to the eye. Neither architecture nor science can agree upon which attributes constitute beauty, as there is no universal definition of it. Biomimicry might seem to embody objective aesthetic notions because humans are hardwired to find nature attractive. In bio-inspired design, beauty is something that is cohesive and respectful of others' surroundings. This is because nature provides numerous attractive examples, at all scales. In fact, from the invisible to the visible, nature seamlessly integrates all the parts of a whole, with not only her forms but also through functions, processes, and systems.

"When I'm working on a problem, I never think about beauty. I think only how to solve the problem. But when I have finished, if the solution is not beautiful, I know it is wrong."
                                                    - R. Buckminster Fuller

Novel Practices in the Built Environment: Dynamic, Atmospheric and Active   Is it possible to create an architecture that is responsive and flexible, engaged and adaptive to an equally dynamic environment in which it is integrated, not simply in a stylistic way, but rather in an active, participatory, and even regenerative way? Can we transform our cities from the inflexibility of a few iconic buildings surrounded by nondescript urban fabric into places made of elements that participate in the nourishment and well-being of their inhabitants? Can we avoid the seldom attractive and instead embrace the meaningfully beautiful that will add value to our everyday and work lives and enhance our quality of life?

Despite the countless number of design competitions and investigations (particularly on paper) that have actively responded to environmental stimuli and cues in the last few decades, there are still a limited number of built examples that move us toward a vision of a dynamic, responsive, and sensory charged architecture. Because it is not yet mainstream, adaptive architecture is very hard for the general public and clients to imagine, further limiting its acceptance and proliferation.

Since the 20th century many architects have been intrigued by and have attempted to build environmentally responsive buildings. Yet only a few succeeded. The most recent and perhaps most renowned is Jean Nouvel's Institute du Monde Arab in Paris (1987). The building's south facade is articulated by a continuous Moresque-inspired screen which blurs the reading of the elements, changing the perception of what is typically understood as window and wall, making them indistinguishable and interchangeable. Computer programmed, mechanically activated "irises" respond to the varying sunlight intensity by opening or closing, thus optimizing interior daylight and comfort. From the street the building reads at times as a whole wall and at other

South facade. Jean Nouvel's Institut du Monde Arabe, Paris, 1987. (Photograph courtesy of J.C. Martin)

Iris detail of a single operable panel for the Institut du Monde Arabe. (Photograph courtesy of J.C. Martin)

times as a whole window. This ambiguous perception, along with the technology that allowed the mechanisms to operate the oculars, was a breakthrough in the history of performative architecture. Drastically different from a typical curtain wall façade in which each individual element would aspire to be all window, all transparent, the library screen south façade appears as an uncanny object in the cityscape, acting and reacting to changing lighting conditions by constricting and dilating – analogous to the human iris. The screens of the facade stand proud against the Paris skyline, yet today, most of the mechanical "pupils" are nonfunctioning. While the beauty of the building remains almost intact, its usability and kinetic aspects are now lost. This building has taught that both the engineering and architecture of highly mechanized buildings must be designed to last. Both, in fact, have to deal with the reality of the complex movement while the technology must be developed with long-term ease of maintenance and durability in mind.

The operability of a building may well be the main challenge in the architectural design process today. In fact, the mechanisms must be designed to match the life span of a building. Ultimately, the engineering of highly adaptive building systems determines

Olafur Eliasson, "The Weather Project," Turbine Hall, Tate Modern, London, 2003. (Photograph courtesy of D. Thair)

their success or failure. Largely, at this point in time, experimentation in the area of active façade technology is often realized only in temporary installations. Illustrious case studies of such ephemeral experiments, such as the Crystal Palace by Joseph Paxton (Great Exhibition of 1851), the Blur Building by Diller Scofidio + Renfro (Swiss Expo 2002), the various pavilions of London's Serpentine gallery, and New York's PS1 summer installations, showcase the most avant garde designs and tests of movement, structural reaction, and adaptation to environmental conditions encountered in transient expositions.

Art has helped advance architecture's investigations and suggests many possible ways to move forward, away from the static and mute massing of the current built environment to more active, responsive, atmospheric and responsible practice. As early as the 1970s, the work of artists such as Gordon Matta-Clark, James Turrell, Dan Flavin, Dan Graham, and Robert Smithson has provoked our spatial imaginations with its complex interplay between architecture and light, sky and atmosphere. One of the major exponents of such artistic tradition today is Olafur Eliasson. With his work he helps us to experience space and time phenomenologically.

In his dynamic installation "The Weather Project" in London (2003), he created an artificial climate within the Turbine Hall of the Tate Modern. Throughout the day, mist was released indoors under the heat of a large artificial sun made of hundreds of lamps radiating yellow light. While lying down on the floor the visitors could see their shadows mirrored as tiny black dots in the hall's ceiling. The transfer of atmospheric effects into an interior space provided the spectators with a sensory experience akin to the light and heat that rises every morning and sets every evening, giving rhythm to our lives. With "Your Atmospheric Colour Atlas" shown at the 21st Century Museum of Contemporary Art, in Kanazawa Japan (2009–2010), Eliasson engaged the viewer in an immersive colored environment saturated with fog, forcing the viewer to have to negotiate between actual and perceived spatial realities, as they lost sight of the edges which define the space. By heightening the visitor's awareness of perception, the augmented reality typical of Eliasson's work provides a powerful moment of clarity. The profundity of such phenomena eludes wordy descriptions, registering instead more sharply in direct experience.

Interior rendering for the proposal for the Museum of Contemporary Art in Wroclaw, Poland, 2008. (Image courtesy of Philippe Rahm Architects)

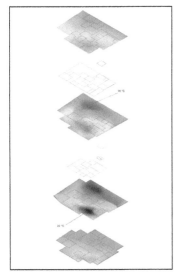

Illustrations generated by climatic software used to model the various climates of the museum to appropriately organize the program. (Image courtesy of Philippe Rahm Architects)

In architecture, some of the most sophisticated phenomenological explorations are found in the work of Philippe Rahm. His interest in affect and performance explores the sensation versus the image, while focusing on meteorological principles to control the interior environment. Rahm's predilection is, in fact, to use gradients of heat and humidity to define space. His entry for the Museum of Contemporary Art in Wroclaw, Poland (2008), combines a deep meteorological and atmospheric understanding of the museum as both a container protecting the arts and a place for contemplation, visitation, and experiential exploration. The program is organized by carefully studying temperature in air stratification: within the cooler environments the majority of the square footage is dedicated to areas of less occupancy (16°C), while the galleries and offices (22°C) are located at the increasingly warmer levels. Driven by climatic requirements the architectural volume emerges in the form of a stepping sectional diagram. The architect's interest in the ephemerality of air and in its "soft solid"[6] state has led him on an investigation focused on "ecology (author's note: rather than program in terms of area and volume) as a driving tool for changing the design process. A process for inventing a new way of living."[7]

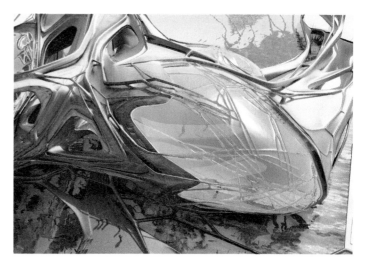

An external rendering of one of the galleries for the Yeosu Oceanic Pavilion. Tom Wiscombe in collaboration with Roland Snooks — Yeosu, South Korea, 2010. (Image courtesy of Tom Wiscombe)

R&Sie's work is singular through their use, manipulation, and distortion of environmental conditions, and construction of dynamic architectures, which result in their mechanistic expression. The team overexposes and almost fetishizes the use of ecological systems to enrich and inform their architectural projects. A significant built example is the house enclosure constructed in 2008, *I'm Lost in Paris*, where the inhabitant assists in the explicit use of the exhibition and the changes occurring in the environment. The hydroponics system nourishes ferns with a bacterial concoction, while the hundreds of glass beakers in which they are contained provide light for the interior spaces. The resultant "living facade" indexes growth, recording environmental changes.

The work of Tom Wiscombe is a passionate exploration of morphology, as relating to the original meaning of the word: the study of natural forms and structures together with a fascination for composite materials. "Ultimately, multi-materiality allows for variable opacity, color and depth effects never before seen in architecture, and possibly only in the transparent head of the Pacific barrel eye fish or the deep colorful organs nested inside the translucent body of the Costa Rican glass frog."[8] Tom's rigorous morphological studies on nature's emergent behaviors find

Interior view of *Hylozoic Ground*. Philip Beesley, Canadian Pavilion at the Venice Biennale, 2010. (Photograph courtesy of I. Mazzoleni)

Detail of the flexible meshwork for *Hylozoic Ground*. (Photograph courtesy of I. Mazzoleni)

their strength in his interdisciplinary collaborative process with biologists, structural engineers, and computational architects. The Yeosu Oceanic Pavilion (2010), done in collaboration with Roland Snooks, exemplifies a new phase of exploration in which the use color as a critical element of communication, beyond its narrow indexical association, is at the center of Wiscombe's interest.

Philip Beesley's work is characterized by his interest in light-weight textile structures that incorporate living organisms. In *Hylozoic Ground*, the installation for the Canadian Pavilion at the Venice Biennale (2010), he explores the use of synthetic pro-tocells, a form of biology that can transform and interact with both the presence of people and the surrounding environmental conditions. This material investigation is often complemented by computational studies in mirroring the way genetics develops its understanding of living organisms, resulting in an immersive experience.

Architects today are also exploring possible future applications of high performance materials capable of self-healing, grow-ing, and decomposing as living organisms. Biomaterials are being developed for biomedicine and defense applications and are inspiring many architects. The most important next steps in architectural advancements will be found in nanotechnologies, materiomics,[9] metamaterials, and biosmart materials' ability to change properties (thermal, luminous, acoustic) or exchange energy (phase change) depending on external environmental conditions.

Materials science, together with emergent technologies, computational and parametric design, find common ground in the complexity of nature and the processes governing it. Nature's fundamental mathematical laws are parametricized with the intention to create performative yet surprisingly bottom-up novel forms.

The iterative nature of methodologies associated with scripting, or coding, often used as form finding, can be successfully applied to any ordering system, including performance-based design. Some professionals prefer to use it in its pure state, algorithmic and parametric only, whereas others use it as the first generative step, to then corrupt it, injecting chaotic and messy elements in the subsequent iterations to provoke distortions, "polluting" the original pureness of the mathematical matrix. In either case, the final moment, the "done" moment, is determined by human determination and likeness of the seen result. While the computational process provides the starting point, only rarely it determines its conclusion. In performance-based design the optimization of energy resources, the limitation of energy consumption, and the optimization of passive systems help shape the end product.

Learning to be attuned to environmental conditions around us, as organisms have adapted to their local environment, is the most deep-rooted lesson we can learn from nature and is the book's inspiration. Animals have anatomically, physiologically and behaviorally adapted to their environment for survival. Our species, with our great intellect, has found ways to bypass the biological course of evolution with buildings, falsely thinking that we can control this divergence and procure only the benefits. While this venture has often failed, bio-inspired design can help us integrate once again with our evolutionary paths. The book focuses on a select number of lessons from nature and suggests some possible solutions. Respectful and responsible steps in design are needed to induce action and change in existing highly anthropomorphized environments without disregarding, but rather regenerating, their

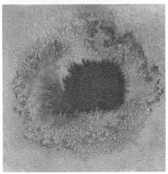

Urban (Dis)solution, charcoal on Strathmore, 2009. (Image courtesy of im studio mi/la with D. Kim)

present conditions. The currently built environment should be seen as something in need of healing. Priority should be given to restoration and transformation, rather than continuing to irresponsibly expand into more — and therefore fewer — rural areas.

Nature provides an infinite source of precedents that have evolved through the years to provide optimal solutions to innumerable challenges of life and survival. Our senses perceive nature's efficient solutions as beauty and our minds are inspired by these elegant natural systems. The analytical brain seeks to understand nature's driving laws and principles, and the creative brain is aesthetically stimulated by these biological creations. Nature thus serves as an endless source of inspiration for those in search of novelty and beauty in efficient and effective design. The transition from bio-inspiration to biomimicry depends on both analytical and creative talents combined with expert knowledge in a conscious effort to synthesize innovative solutions through a persistent search of nature's repertoire.

"Nothing in biology makes sense except in the light of evolution."
- Theodosius Dobzhansky

How Biology Informs Architecture

Biology is the study of life. Defining life is challenging, and there is no absolute consensus for its definition. Most scientists agree, however, that several traits characterize all living things. Living organisms undergo chemical processes for energy and maintenance. Cells gain energy through a constant cycle of breaking down matter and constructing matter into cellular components. Living beings reproduce, either sexually or asexually. They are able to evolve and adapt to their environment over time, so that characteristics of a species change from generation to generation. They respond to stimuli, and in doing so, they maintain homeostasis. Life is dynamic, meaning that all traits that characterize living beings involve dynamic processes.

The basic unit of life is the cell; it is the building block of all life forms. A living organism can consist of only one cell or of multiple differentiated cells. Though single-celled organisms, like most bacteria, seem very different from organisms with millions of cells, they are still connected, as each undergoes all of the basic processes of life.

The field of biology seeks to understand the patterns and processes that govern all living things. Biological patterns and processes are incredibly diverse and complex, and many disciplines of biology exist to understand how life works. For example, biochemistry is the study of chemical processes fundamental to living organisms such as metabolism and cell signaling. The study of the components of cells, how they work, and how they interact with their environment is the focus of cell biology. The discipline of genetics is concerned with gene function and how genetic information gets passed from one generation to the next. On a larger scale, evolutionary biology seeks to understand how species originate and how they change over time, while ecology is concerned with broad scale patterns of interactions among organisms and interactions among organisms and their environment. Biology is also interdisciplinary; it integrates with other fields. Biomathematics uses models to understand and predict life

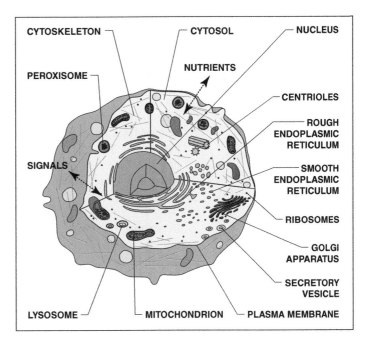

Diagram of a generalized animal cell. (Image courtesy of im studio mi/la)

processes. Bioinformatics uses information technology to store and analyze biological data. Bioengineering uses biological principles for application in design and manufacturing.

Scientists employ a specific methodology, called the scientific method, to acquire knowledge about the natural world. This methodology is the basis for all modern scientific research, as it is evidence based and strives to be objective. The process involves inductive reasoning, moving from observations of the world to deriving general patterns and principles about how the world works. First, observations are made of a phenomenon in the natural world. Then, a hypothesis is formulated to explain those observations. Predictions are made that follow from the hypothesis. In science a hypothesis must be testable. It needs to be framed so that some kind of measurements can be taken in order to determine that the hypothesis is true. A hypothesis must also be falsifiable; scientists must be able to show that it is incorrect. Consequently, an observational or experimental study can be designed with the goal of testing the hypothesis, and the results from the study are used to make an interpretation of the hypothesis. The data will either support or reject the hypothesis. Through multiple, sequential testing of related hypotheses scientific knowledge is gained. The scientific method is necessarily iterative. An idea has to be tested over and over again before it is accepted, and even then, new evidence may come to light that refutes previously supported hypotheses. Therefore, ideas in science are always changing. The process of conducting science is a repeatable, ever-refinable one, where hypotheses are constantly tested and nothing is ever proven absolutely.

In contrast, the design or creative process differs from the scientific method and is more a deductive reasoning process, beginning with the knowledge and study of general principles which in turn lead to specific solutions. Building design emerges in response to the rise of contextual questions and the necessity for a specific population to provide for needs related mainly to function, protection and comfort.

The projects go through several steps, from the conceptual to the detail of building components. While the conceptual and schematic phases are characterized by their intuitive, open, even "messy" process, decisions become more definitive with each iterative step and less able to change as the design approaches the construction phase. This contrasts with the scientific method, which allows for continuous change to be taken into consideration and tested. The early creative process, especially in its beginning phase, is of specific interest in this book. Architecture looks for precedence to confirm, validate, but most importantly expand, its design exploration in search of innovative ideas. Nature itself is offering us a meaningful source of new precedents for design.

Humans are understandably driven to explore the world around us, and for good reason. Throughout the evolutionary history of our own species, biological knowledge has been imperative for survival. Humans needed to learn which plants were safe to eat and which ones were poisonous. It was crucial to understand the behavioral patterns of predators to avoid becoming prey. In one form or another humans have gained biological knowledge over millennia. However, despite the number of biological disciplines and the amount of time we have been studying natural phenomena, we still know relatively little about the natural world. There are vast numbers of species, biological processes and biological interactions that are yet to be understood. We do not completely understand how life evolved, how many species exist on Earth, or the details of how the genetic code operates. These are just a few of the fundamental questions that biology still has to answer.

The disciplines of design and architecture can extrapolate knowledge from the biological world in order to improve the way humans live. The field of biomimicry applies biological principles to design in two main ways. First, a focus of biomimicry is to understand the dynamic context within which we operate and place buildings into the environment. Study of the natural world takes into consideration the interconnectedness and diversity of nature, and architects can learn from this perspective in order to design elements that can become integral parts of natural systems. Biomimetic designers strive to be more considerate of the environment, to be less invasive and more conscious about the fact that humans live among other components of nature. Second, designers can learn about the functional aspects, or adaptations, of organisms and translate those principles into design. Biological form is studied for its function, which is a basic precept that can be imbued into architectural design through the study of precedents. The starting point for such investigation can occur at either large design scale, such as that of a city or neighborhood, looking for inspiration in the complexity and interrelatedness of ecosystems, or by choosing one element of an animal as focus of inspiration, for example, starting with an animal skin to use as model for the design of a building envelope. In this book we develop both principles, recognizing that nature teaches us that everything is connected; therefore, our study of the building envelope is not meant as design of a part in isolation, but rather the opposite, which focuses on that part in the context of the whole.

Evolving and Adapting to Survive   Life on Earth has changed over time. Living beings appeared approximately 3.5 billion years ago in the form of single-celled, aquatic relatives of bacteria. Since then, millions of species have come into existence and have taken myriad forms. This process is called evolution and is the unifying theory of biology. Evolution is the conceptual framework that brings together all the disciplines of biology, because it explains how living things came to be, how they are related to each other, and how they function. In its most basic form, evolution simply

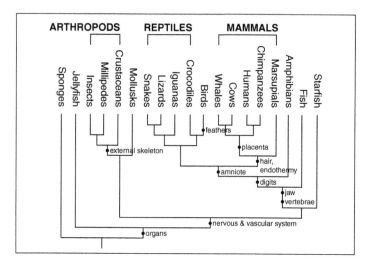

An evolutionary tree, or phylogeny, of animals. Major evolutionary adaptations are depicted on the tree. (Image courtesy of im studio mi/la)

means change in the characteristics of a population over time. For organisms with short generation times, this process is directly observable. All living beings share the same genetic code, and evolution occurs when the frequencies of different forms of genes change over time.

Four major mechanisms of evolution are recognized: mutation, natural selection, genetic drift, and gene flow. Mutation is the basis for all genetic variation. It is the spontaneous change of genetic material, which can be caused by external factors such as radiation or chemicals, or by internal, genetic processes, such as errors that occur when DNA is replicated. Genetic variation is the underpinning for all other processes of evolution. It must be present in order for evolution to occur.

The mechanism of natural selection was a revolutionary theory proposed by the English naturalist Charles Darwin (1809–1882), and it explains how organisms adapt to their environment. The theory of natural selection states that individuals within a population have variation in their traits. Because of the inherent variation in populations, some individuals will have traits that are better suited to the environment than others. Individuals possessing those traits will survive and reproduce more than others in the population. If the traits are heritable, that is, if the traits have a genetic basis, then this process leads to change in the characteristics of populations over generations because individuals with the most well-adapted traits pass them on to their offspring. An important prediction of this process is that the traits being selected for are only those that allow organisms to survive in the environment that they live in at the time. When the environment changes, which it does continuously, the traits that were previously successful may not be adapted to the new environment. If there is enough variation, the characteristics of the population may change again. If not, then the population may become extinct.

The other two mechanisms of evolution do not actively help populations adapt to their environment, but do change the frequencies of genes in a population. Genetic drift occurs when gene frequencies change due to random sampling. The frequency of traits in a population that do not necessarily have an adaptive value can vary over time based on which individuals happen to mate and pass on traits to the next generation. Gene flow occurs when individuals from one population migrate to other populations. This migration tends to add genetic variation into a population as well as alter the frequencies of forms of a gene in populations.

Biologists recognize two major scales of evolution, which in reality are part of the same continuum. Macroevolution refers to processes at or above the species level. Studying speciation, the process whereby new species evolve, or how traits of organisms have changed over millennia using the fossil record are examples of macroevolutionary research. Most people think of evolution at this scale. However, evolution can occur over smaller time frames as well. Evolution within a population or a species is referred to as microevolution. The frequencies of forms of a gene are usually constantly changing within populations, even though they may not yield large scale, obvious changes and do not give rise to new species. The development of antibiotic resistance in bacteria is a prime example of microevolution.

On the macroevolutionary scale, key innovations have led to the success of many groups of animals. Key innovations are traits or characteristics of a group of organisms that allow it to diversify and give rise to many species. These traits may give organisms

(Top) The origin of the eye is thought to be a key innovation in the evolution of animals. (Photograph courtesy of N. Amin)

(Bottom) Coevolution is common between plants and their pollinators, such as bees. The plants have their pollen spread to other plants, and, in turn, pollinators get rewards such as nectar. (Photograph courtesy of S. McCann)

the ability to acquire more resources or they may allow them to exploit previously unavailable resources, promoting expansion into new habitats or niches. There are many examples of key innovations, and they range in scale from the evolution of wings and flight to the evolution of cusps on teeth that allow for the use of new dietary resources. The evolution of the eye exemplifies a particularly successful key innovation, because so many more animals possess eyes than do not possess them.

The evolutionary interplay between predators and prey is often interpreted in terms of coevolution. Coevolution is defined as reciprocal evolutionary change between species; an evolutionary change in one species causes one in another species. Animals do not exist in a vacuum. They interact with many other organisms in their ecosystem, and these work to influence evolution in each other. Besides the coevolution between predators and prey, another classic coevolutionary example is between parasites and their hosts. Coevolution is an important aspect of evolutionary study given that it is likely that every living species on the planet has coevolved with other species.

The concept of adaptation to the environment has not traditionally been central to architectural design, yet to improve our ecological footprint it needs to be taken into consideration. The pace of growth of human building construction worldwide is more rapid than the rate of evolution operating on biological timescales. Having more buildings equates to less animal habitat, less resources and more pollution. Slowing the pace of building by humans while being cognizant of how architecture can fit into the environment around it will help achieve the goal of lessening and perhaps improving our negative impact on the Earth.

Climate and Biomes Throughout our history humans have caused major, negative impacts to the planet. As technology has progressed and our population has grown, the rate of human-induced changes has rapidly increased. Much of what has driven these negative effects on the environment is the absence of being willing to change even given the harm we know we are causing. A major issue facing the planet now is global climate change. The increase of carbon dioxide, primarily due to the burning of fossil fuels, together with the release of other greenhouse gases into the atmosphere, is having significant impacts on regional and global climate patterns.

Because of the spherical shape of the Earth the sun's rays are not equally spread across the planet. Areas around the equator receive the most direct light, which is why they are warm. In contrast, areas at higher latitudes are cooler, because radiation from the sun is spread over a larger area. (Image courtesy of im studio mi/la)

Each terrestrial region in the world is characterized by major vegetation types called biomes, and each biome is associated with a particular set of climatic conditions. The term "climate" is used to refer to a suite of variables such as temperature, moisture, sunlight and wind that characterize a region when considered over long periods of time. Weather is also sometimes used to refer to the same set of variables, but over much shorter time scales. Large-scale climate patterns are regulated by the Earth's shape, tilt on its axis, and orbit around the sun. Global temperature patterns are governed by the angle at which the sun's rays hit different parts of the Earth. The spherical shape of the Earth means that the equator receives more direct solar radiation per unit area than regions closer to the poles; therefore, equatorial regions are warm. Higher latitudes are cooler, because the same amount of sunlight is spread over a larger area and has traveled farther to reach the Earth's surface.

Global patterns of precipitation are influenced by cycles of air circulation. Areas along the equator receive the most moisture, while areas at around 30° latitude north and south are the driest on Earth. This is due to the differing densities of warm and cold air. Warm air is less dense than cold air, causing warm air to rise.

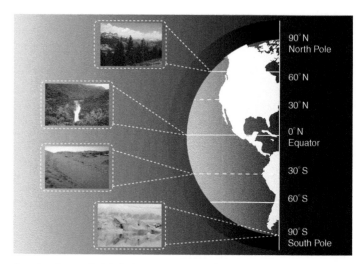

A major driver of global climate patterns is the way in which air circulates latitudinally. Cycles of evaporation and condensation create areas with moisture around the equator and at 60° latitude, while latitudes of 30° and 90° are dry. (Image courtesy of im studio mi/la)

When the sunlight hits the equator it causes warming of the air as well as evaporation of water. This warm, moisture-laden air rises, cools and condenses in the atmosphere, and then falls as precipitation. Because the Earth spins on its axis, rotating air currents move air in a north to south pattern, creating cycles of evaporation, condensation and precipitation. As the warm air moves toward the poles it cools, and at around 30° latitude, begins to descend. This cool dry air mass absorbs moisture from the Earth's surface, creating arid conditions at that latitude. At latitudes around 60° the cycle repeats; air rises, cools and releases precipitation. The cold, dry air then travels toward the poles, where it absorbs moisture and creates cold, dry tundra.

Seasons, annual fluctuations in temperature and precipitation, are caused by the Earth's 23.5° tilt on its axis. Summer occurs when either the northern or southern hemisphere is most tilted toward the sun as the Earth circles around it, and winter occurs when each hemisphere is tilted away from the sun. Regions near the equator do not experience major changes in temperature, but precipitation does vary.

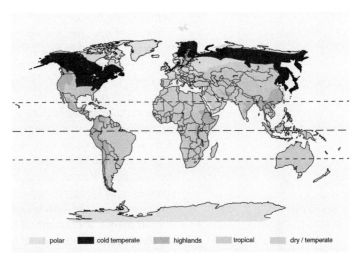

The Köppen classification system, one of the most widely used systems, divides global climate into five major types. (Image courtesy of im studio mi/la)

Ocean currents, which determine water temperature and greatly affect air temperature, are influenced by air circulation patterns, the rotation of the Earth, and the intensity of sunlight hitting the Earth's surface. Warm waters at the equator are carried toward the polar regions along the eastern coasts of the continents, while cold water traveling toward the equator moves along the west coast of continents. These patterns lead to differences in the distribution of warm and cold water in the oceans, as well as changes in the distribution of warm and cold air on the Earth's surface as it travels with the water.

The most common climate classification scheme is called the Köppen system, which defines five broad climate types: tropical, dry, highlands, cold temperate, and polar. These types can be further divided into secondary classifications. These secondary classifications often correspond to major biomes, depending on classification scheme.

Biomes tend to have similar communities of animals within them, because the animals have evolved in similar climatic conditions. Climate, however, has changed over time, and accordingly the distribution of animals on Earth has changed with it.

The field of biogeography focuses on understanding the distribution of species in space and time. The location of species is a result of current and past climate. The Earth's climate has changed over millennia due to slight but continuous changes in the shape of the orbit of the Earth around the sun, the tilt on its axis, and in the orientation within its orbit. Over geologic time the amount of sunlight hitting the surface of the Earth and how air and water circulate are altered. Subsequently, warm periods that can last for thousands of years alternate with cool periods, often accompanied by glaciation. Additionally, the movement of the continents has shifted significantly over time, which causes changes in biogeographic patterns. The Earth's crust is divided into large, thick (50–250 miles) plates that constantly move over the soft mantle underneath. It has been hypothesized that a giant supercontinent existed about 200 million years ago that broke up over time, leading to the current position of continents. This process of continental drift explains many patterns in plant and animal communities over the Earth. Closely related organisms can occur in very different places if they evolved on a land mass that later split apart. Conversely, unrelated animals can have similar sets of adaptations if they have evolved in similar climatic regimes.

Ecosystems and Biodiversity   Over the last 3.5 billion years, a remarkable number of species evolved. Almost two million species have been named, and it is likely that millions more have yet to be discovered. Biodiversity, or the variation of all life forms, can represent different levels of biological organization. Genetic, ecosystem and species diversity are all considered part of biodiversity. Most commonly, though, biodiversity refers to the number and composition of species in a given place, and maintaining biodiversity is incredibly important for ecosystem stability and function. Ecosystems are all the co-occurring organisms and abiotic conditions in a particular area, and they function as an integrated unit. Many ecosystems occur within a biome. Some ecosystems have incredibly high levels of biodiversity, such as rainforests, while others, such as the tundra, have sparse numbers of species.

Estimates of undiscovered species range from five million to fifty million. In general, species that are large and charismatic have been named and studied significantly more than small or microscopic species, such as invertebrates, fungi, and bacteria.

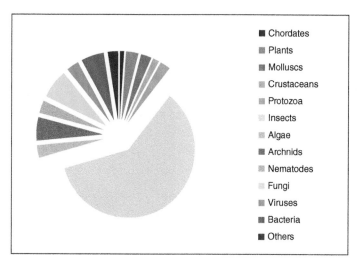

A pie chart representing estimated species richness for major groups of organisms. Data from Purvis and Hector (2000). Image adapted by im studio mi/la. (Image data courtesy of Nature Publishing).

Approximately 15,000 unknown species are identified each year, most of them being from less well-known and studied groups. Though most people think of plants and animals when they think of biodiversity, the greatest amount of diversity is composed of single-celled organisms. These organisms are incredibly diverse in how they live. Some use oxygen for respiration, some convert light energy to carbon compounds, and some even use inorganic chemical compounds for energy.

Because they function as a unit, the processes that govern ecosystems include both random, or stochastic, events such as natural disasters, as well as the more predictable interactions and feedback among organisms, such as predation, competition, parasitism, and mutualism. The outcome of interactions among species can be destructive to both interacting organisms, as in the case of competitive interactions for resources. Alternatively, the outcome may be positive, as in a mutualism, which is a symbiosis that benefits individuals of both interacting species. Finally, an interaction can have a negative impact on one species but a positive impact on another, as in the case of predation where one organism is eaten but the other gets nutrition to sustain life.

Neotropical rainforest ecosystems, such as this lowland forest in Ecuador, hold extremely high levels of biodiversity. (Photograph courtesy of S. Price)

The loss of biodiversity and ecosystem function induced by human activities is occurring at an alarming rate. Dozens of species are likely going extinct each day, many without ever having been discovered. It has been estimated that current extinction rates are 100 to 1000 times greater than baseline level extinction rates over geologic time. This current level of extinction is comparable to those documented during mass extinction events, such as the Cretaceous-Tertiary extinction event that killed the dinosaurs, during which 75% or more of species become extinct over a short period of geological time. There are many causes of species loss; chief among them are habitat destruction, climate change and the invasion of non-native species.

Ecosystems provide many services and resources for humans, and the loss of species within them can have negative impacts for the human population. We extract wood and food from ecosystems, we rely on them to recycle nutrients, plants absorb the carbon dioxide that humans produce, and they help regulate climate. Conservation biologists Paul and Anne Erlich have warned that the health of ecosystems depends on all of the biological diversity within them; the removal of any one species has some measurable, negative impact on the system and the removal of more and more species will at some point break the system. Consequently, ecosystem function will be lost. In the face of human-mediated environmental change, conserving higher levels of biodiversity can provide greater resilience against ecosystem collapse and mass extinction.

Biomimicry seeks to link complexity, biodiversity and coexistence, using them as precedents for principles of design where, even when the starting point of the design investigation involves a singular element, its development is understood as part of a whole system that interrelates natural and man-made components, allowing for a beneficial coexistence.

"Humans have a tendency to fall prey to the illusion that their economy is at the very center of the universe, forgetting that the biosphere is what ultimately sustains all systems, both man-made and natural. In this sense, 'environmental issues' are not about saving the planet—it will always survive and evolve with new combinations of atom (sic)—but about the prosperous development of our own species."

- Carl Folke

Lessons Learned: Biology to the Built Environment

All humans have an inherent connection to nature on some level. Our innate ideas of beauty are connected to forms in the natural world. Humans feel an instinctive, primal draw to nature because we have been very closely connected to it for the entire time we have existed on Earth. The desire to connect to nature is ever-present, and this longing provides an entry point for architecture.

Recent advances in science and technology allow for more elements of the natural world to be used as inspiration. There are so many facets of the living world to study – from ecosystems to microscopic organisms and even DNA. There are big pushes to advance scientific innovation and creativity as well as the globalization of information. Information can be transferred across the world almost instantaneously, and interdisciplinary work is facilitated by these advancements.

Architecture has a lot to learn from science, both from a methodological perspective, as a source of design precedents, and as a springboard for new ideas. As new research is developed in the fundamental as well as applied sciences, architecture can increasingly find models for bio-inspired design through applied concepts and the discovery of new study tools. For example, the materials sciences are entering an exciting new phase related to multi-scale material systems – materiomics – which looks at not only the incredibly small, but also at how observations can be made in an interrelated way across different scales.

The method of investigation presented in Part II starts with one element of biology, the skin, and one element of architecture,

Polar bear, *Ursus maritimus*.
(Photograph courtesy of M. Johnson)

the building envelope, yet these components also serve as larger ideas, as bridges between realms. The skin and building envelopes exemplify the dialogue between internal and external worlds as they respond to inputs from both. They protect us from the environment while keeping us connected to it. The skin is just one of innumerable aspects of biology that can serve as inspiration for design, but the use of skin and building envelopes here clearly defined a starting point for our investigation and vision.

Our exploration of skins and building envelopes unfolds through a marriage of methodologies – combining the linear and analytical scientific method with the more lateral, synthetic, deductive creative design process. The difference between these processes enhances the potential for design. Scientific ideas are never certain and can never be proved absolutely. The search for more and more support for scientific hypotheses is never-ending. When thinking about bio-inspired design, we need to consider how its success is measured. In science, if a hypothesis is shown to be incorrect, it is discarded. In design, the measure of success has to do with whether the idea is a springboard for further inspiration and for what emerges in the design realm. In design the designer defines the goals. If the goals are met, the design is successful. In science the objective is to understand nature to its fullest extent. In architecture, nature is used to inspire design and its processes whether or not the underlying theories about nature remain valid through time. If the resulting design concepts that come from biological observation in architecture are innovative and move design thought forward, it does not matter if the science is later proven wrong. The determinant of the project's success is its function. In this book, we offer two case studies (in Part II), each with a particular story.

The polar bear project, the first animal studied for this book, was developed prior to the publication of a paper correcting a previous understanding of how this animal's hollow hair follicles might be contributing to the thermoregulation of the animal. The new paper demonstrated through colored scanning electron micrography how the internal walls of the hollow guard hair could not conduct heat. This new knowledge of fur morphology does not compromise the architecture developed from the original idea because its functional performance as designed is still valid. The active building envelope is covered with tube-like structures that use a mechanized tracking system to follow the movement of the

Namib desert beetle, *Physasterna cribiripes*. (Photograph courtesy of A. Sosio)

sun, maximizing heat absorption by the thick insulated envelope. This mechanism is still valid, even if the science that inspired it is no longer found to be true.

Another case study involves Namib desert beetles and their water harvesting strategies. One particular species was studied for its ability to collect water through a behavior known as fog basking. The study concluded that bumps on the dorsal surface of the beetle have physical properties that allow water droplets to be collected and roll into the beetle's mouth. It was later discovered that the species the study focused on was incorrectly identified, and it was not even based on a species that fog basks. The discovery happened during the development of this book and led to the project being iteratively revisited, adjusted and rearticulated based on this new scientific development. Each project's inspiration follows its own path, and what emerges from this is the importance of the continued communication between science and design. We attempt to teach an investigative interdisciplinary methodology intended to provide new precedents for study while permitting creative minds to make the leaps and lateral moves, necessary to the design process.

The ability to understand the interrelation of systems is important to all design professionals and can be enhanced by interdisciplinary education. Seeing everything as connected is an ability not common to all, but something that can be acquired. Natural sciences are a great place where a designer can learn about connectivity despite some scientists' tendencies to focus on detail. It is not the discipline but the individual's mind-set that matters. This process happens through use of an interdisciplinary team where each specialized member keeps an open mind toward discovery and translation. The ultimate goal of our studies in this book is not only to apply the organism's features to design, but to gain meaningful inspiration from the way in which that application enhances creativity in the design process. It is not only the specific functional lessons that an organism conveys but the broader themes of interrelatedness between components of nature that can be applied to the architecture. Our way of working emerges from a conscious collaborative effort to observe and then find design concepts from two worlds that initially seem to have little in common – the built environment and nature.

While Part I discusses the biological and architectural background, Part II describes our methodology, which is applied to twelve case studies. In particular, Part II looks at how architecture is inspired by four key biological functions: communication, thermoregulation, water balance, and protection. Through initial biological analysis and the development of the projects, the proto-architectural proposals aim to provide a pivotal point in the relation between the built and un-built environment, providing a shift from the exploitation of nature to exploration and collaboration with it.

PART II

2 Applications

"Consciously emulating Nature's genius means viewing and valuing the natural world differently. In biomimicry, we look at Nature as model, mentor, and measure."

- J.M. Benyus

Overview

Our approach to biomimicry seeks to expand on the most widely used applications of the discipline to design and architecture. Rather than mimicking or recreating nature through design, we draw primarily from the functional, or performative, aspects of nature and integrate those with the environmental context in which such aspects are found. The natural world informs design through a synthetic understanding of the adaptations an organism has to its environment and the relationship it has with other organisms in the ecosystem in which it is a part. Using this approach we can bring functionality to designed elements and thereby "inform the form." The resulting project is not a direct translation from a particular organism, but is inspired by study of the function and its context within a natural system. Novel and unexpected forms consequently emerge from this explorative process. Though the primary driver of the design is function, form is certainly not forgotten as it constitutes a fundamental aspect of design. Humans are instinctively drawn to the complex, diverse and elegant forms from the natural world. Through these projects, by beginning the process with the performative aspects of nature, we attempt to ultimately achieve beauty. In this way, nature is a great model.

Bio-inspired design can help achieve the goal of lessening our impact on the environment. Rather than creating a distinction between the built and un-built worlds, our approach interrelates the two, recognizing that all systems in the natural world are continuous and coexist with each other. Rather than inserting architecture into nature, our goal is to shift this perception and integrate built forms into the natural world. The merging of biology and design leads to discovery, innovation, and a novel position in our relationship to the environment. This paradigm shift will only occur by initiating a dialogue in contemporary architecture that moves beyond pure formal and sustainability concerns and aspires to connect directly with nature on a performative level. Our proposed method offers one way of exploring biomimicry through functional analysis, and allied disciplines are taking a similar approach. Parametric design, for example, has renewed interest in the complexity of nature and processes governing natural phenomena. Advanced technologies look to the natural world to help modify how products are manufactured and provide comfort to human habitations. As society moves toward an environmental consciousness, we are encouraging the use and understanding of nature to reformulate our aesthetic sense. To illustrate our methodology we have used examples from the animal kingdom as a source of inspiration. We considered the climatic and ecological context in which animals evolved, and we drew upon those adaptations that have allowed them to become successful in those contexts. Specifically, we focused on animal skins, translating the observed adaptations into the built environment. Skin offers one of the major models or sources of inspiration that architecture can draw from biology. There is an abundance of other models, and the book's method should be used as a way to guide oneself through the process in an analogous fashion. We use the term "skin" on a general level to refer to any animal covering, including fur, feathers, scales, exoskeletons, and shells. Animals possess remarkable variation in the type of skin they have and in the ways they use the critical functions of skin, such as protection, sensation, and heat and water regulation. To draw an architectural parallel, building envelopes serve multiple roles, as interfaces between building inhabitants and environmental elements (e.g., water, air, sound, light and temperature). That is why the word "skin" is often used to refer to building envelopes. However, while recognizing the parallels between natural skin and building envelopes, we strive to avoid simplistic interpretations, such as attributing anatomical characteristics to buildings.

The work presented here is a series of proto-architectural projects that focus on the development of the architectural envelope. These projects explore the essential meaning of a building and its envelope as they investigate the elements that provide shelter, protecting inhabitants from external forces while creating an interface with the environment. The resulting projects perform and respond; they take into consideration the dynamic local environmental conditions, enhancing and supporting these conditions rather than exploiting them. Consequently, the focus on building envelopes greatly facilitates the creation of a more sustainable way of building and living.

While animal skins constitute the main driver for the design inspiration, the studies show recognition of the importance of a systems integrated approach. Once the skin's major performative function was explored, we expanded our research to a physiological and behavioral analysis of adaptations that help an organism succeed and survive. This examination led to other elements important to the architectural resolution of the design. The process starts with the fundamental boundary element, the skin, with the understanding that in order to design a coherent envelope, which in nature and architecture comprises a system, it is important to describe the elements, connections and interactions between the components of that system. Animal skins and building envelopes are a departing point for our approach to bio-inspired design. As boundaries, they provide an essential element for connection and integration of inside and outside, of built and unbuilt.

The following chapters are organized around four important skin functions: communication, thermoregulation, water balance and protection. Each chapter has two parts. The first introduces and analyzes the selected function, illustrated using animal examples. The second presents a few proto-architectural projects based on in-depth analyses of individual animal skins and associated systems.

Methodology

The book draws examples from the animal world that exhibit the four major functions of skin we have chosen to examine. Our thesis is that through an evaluation of how the environment influences the anatomical, behavioral and physiological adaptations of animals, we can develop a deeper knowledge of how human constructions are affected by, and affect, the environment. This process results in an increased design intelligence derived from studying a range of investigative opportunities selected from the observation of nature. Only limited components of both the animal and the design are shown, allowing for an openness of interpretation and exploration of the remaining aspects.

Rigorous and systematic analyses of the performative aspects of the animal skin are the initial steps of the proposed methodology. Climate and habitat data are taken from the scientific literature, and the organism is analyzed in its environmental context. Subsequently, drawings based on observations of the animal are made in order to formulate possible architectural applications. The analysis of the animal is conducted through line drawings, diagrams, and renderings of the organisms to yield a synthetic yet distilled portrait of the architecturally relevant adaptations. The architectural rendition of biology is simplified and selective as it draws from biological solutions to environmental challenges in nature and describes them in a language that traditionally pertains to architecture and design.

As the proto-architectural project is developed, the scientific process is combined with the creative process. Though the scientific process requires creativity, it tends to function in a more stepwise and iterative manner, while the creative process is characterized by leaps and returns; it is labyrinth-like and can appear, in fact, quite "messy." We allow for and encourage this nonlinearity, while simultaneously ensuring that the design is still fundamentally based on a platform of biological information. During the proto-architectural exploration, the bio-inspired research is consistently referred to, and the biological concept is present in the ultimate design. The proposed methodology enhances and strengthens the links between the biological and the architectural components.

The challenge likely to be encountered with this truly interdisciplinary methodology is the necessity to build a common linguistic platform for communication. It is important to recognize the need for the time it takes to develop a shared vocabulary and tools to lead to effective comprehension and information exchange and reduce the semantic noises common to such work. This process will be different for each team and each task, and it is likely that a common language needs to be developed for each goal. In our case the challenge has been greatly facilitated through drawings. They help identify, evaluate and define the knowledge relevant to the overall objectives. The generation of a common visual language has allowed for enhanced opportunities for innovative solutions to be created between disciplines.

Both the biological elements and architectural ideas are explored using the most diverse visual tools available. First, line drawings are created to isolate key elements by indicating the functions that are being examined. Taking the animal's body as a starting point, the drawings grow step by step like architectural plans and sections, retaining the scale of the original and concentrating on one part of the body to define it in detail. The aim is thus to create a stratigraphic sectional exploration of the skin, the primary organ of investigation and design inspiration. The diversity of tools ranges from hand-sketched drawings to 3-D digital modeling, physical models, and digitally fabricated mock-ups. A process of initial research leads to a codified design through a process of graphical analysis and design synthesis. The final rendition of the designs are proto-architectural projects rendered to substantiate the final conceptual stage.

The projects explore the insertion of the design element in a particular setting so as to arrive at the material detailed definition of a wall section, which aspires to describe the material components and their performative characteristics. The basis for the design might contemplate one specific animal trait or might emerge through the synthesis of multiple traits. At times the propositions that develop may not directly coincide with the initial feature analyzed; however, they potentially engage and result from its form or function applied to other aspects of the design. A building envelope is often constituted of multiple parts that are either layered or otherwise assembled to provide the functional qualities required. These parts are interrelated, similarly to the way in which the skin interrelates with other systems in living organisms; the way they are connected implies a multiplicity of functions that are in balance.

The following twelve projects exemplify the conceptual framework outlined here. Each design project is organized into ten parts (pages) to explain the biological research and inspiration and subsequent development of the architectural project.

Project Parts

Animal Examples   The animals introduced on this page share similar characteristics with the focal project species. The examples listed are then used to further demonstrate one of the four major skin functions.

Taxonomy   Each project focuses on one species due to the qualities it exemplifies in relation to the chosen skin function.

Habitat & Climate   The habitat, climate and ecological context of the animal species are described. Selected variables are detailed, including temperature, precipitation, humidity and wind patterns. This information is particularly relevant as an introductory element to the biological analysis as well as the architectural proposition.

Animal Physiological, Behavioral & Anatomical Elements
Major adaptations applicable to the four skin functions under study are introduced with structural, physiological and behavioral traits highlighted.

Interface between the Skin & External World   The type of skin covering is described as well as the components that comprise the skin. Aspects of the skin that exemplify the functions being studied are shown.

Interrelationship between the Skin & Internal Systems   A description of the internal physiological systems that interface with the skin is given. The systems were chosen to complement the major skin attribute of the species chosen.

Proto-Architectural Project   The design program and location are introduced. The main design drivers and strategies are shown through diagrams that describe environmental forces and scenarios (for example, day/night, summer/winter, and rainy/dry seasons) critical to the understanding of the envelope's development. The location chosen is within the habitat of the animal. Assignments were given in a didactic manner to facilitate the design process.

Project Documentation   Diagrams and details derived from the biomimetic aspects of the design are rendered, such as modular, tessellating, and aggregating strategies.

Section of the Building Envelope   A wall section shows the articulation of the key project components and describes the entire wall assembly in its performative and material aspects.

Project 3-D Rendering   The final rendering offers an atmospheric and phenomenological overview of the project.

Skin Composition and Functions   The diversity of animal skins provides fertile ground for bio-inspired design. Animals have adapted to a wide range of environments, including those extreme in temperature and rainfall, and animal groups have unique strategies for dealing with the climate they exist in.

The skin is an ideal organ to use as inspiration in architecture because of its multifaceted functions. It performs multiple, complex tasks yet it is one definable and readily visible system of the body. Skin is also a duality. It constitutes the threshold between the interior and exterior realms; it is the element of connection between the two. Skin is a barrier, yet it is permeable. These comparisons are true and apply to both the biological and the architectural worlds. Using skin with its many parts and functions as inspiration for the design of building envelopes allows us take a holistic and systems-based approach to the proposed projects.

The skin performs numerous critical functions. It acts as a protective barrier against pathogens, predators and the environment. It aids in temperature regulation through mediation of heat gain and loss. Skin regulates water balance in the body through preventing its loss and by storing it or releasing it. It is a permeable barrier that allows essential elements like oxygen and nitrogen to diffuse into it. Nerve endings that respond to temperature and

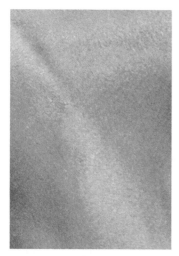

Close-up of human skin. (Photograph courtesy of S. Yelisee)

pressure are contained in the skin, leading to changes in sensation. The patterning and coloration of skin serves important communication functions. Through all of these diverse functions, the skin can serve as a starting point for innovation, expansion and enhancement in architectural design.

Skin is technically the soft outer covering of animals. However, in this book the term is used in a much broader sense to mean any animal covering. Examples of organisms with fur, feathers and scales are used, because these are structures that originate from the skin. Any structure that encloses an animal's body and serves protective and regulatory functions provides a point of departure for our explorations.

To introduce details of the structure, function and physiology of skin we have chosen to describe human skin, because it provides an easily understood model.[10] Human skin is composed of three major layers: the epidermis, the dermis, and the hypodermis. The epidermis is a thin surface layer that acts primarily as protection. It contains keratin, a protein, which keeps water in and harmful chemicals and pathogens out. Specialized cells called Langerhans cells engulf invading microorganisms and send messages to the immune system for activation. Melanin, a pigment, is produced in the epidermis. This pigment proliferates in reaction to UV light and protects deeper tissue from sun damage (the process of tanning). The epidermis also functions in preliminary vitamin D production, where chemicals produced interact with UV light, leading to remote formation of the molecule that promotes calcium absorption. Too little vitamin D can weaken bones, and thus this process is critical for human health.

The layer beneath the epidermis is called the dermis. It functions in blood circulation, thermoregulation, protection from stress and strain, and sensation. The dermis allows blood vessels located in the lower skin layer to bring oxygen, water and nutrients to the epidermis, where growing cells are fed through diffusion and osmosis. When the body is hot, blood vessels and capillaries dilate and conduct heat to the epidermis for surface cooling through radiation and sweat production. When cold, vessels constrict and retain body heat. In addition, nerve endings and special receptors are located in the dermis that allow the body to respond to heat and touch sensations. A thick layer of tissue within the dermis has collagen and elastic fibers, providing insulation, cushioning and

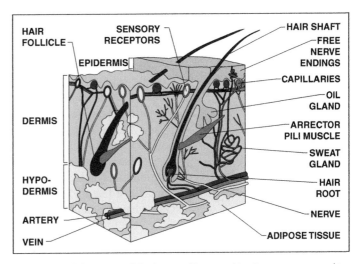

A generalized depiction of the layers of human skin. (Image courtesy of im studio mi/la)

elasticity. Hair follicles and sweat glands are also located in this stratum.

Below the dermis the hypodermis is found. It is composed of a thick layer of loose connective tissue containing larger numbers of adipose, or fat, cells, which store energy. Major blood vessels also occur in this layer. The hypodermal layer provides insulation, acts as a shock absorber, and allows the skin to slide smoothly over muscles, bones and joints.

As well as integrating with the major internal body systems, such as the circulatory and nervous systems the skin also interrelates with the digestive system. The digestive system is crucial for skin function as the intake and digestion of fats and essential oils facilitate and maintain the skin's function as a protective barrier, as well as nourish glands and hair follicles. Vitamin D production by the skin helps the uptake of calcium from food introduced into the alimentary (or gastrointestinal) system. In the nervous system the skin acts as the initial source of input, the signals of which are transferred to the brain, for senses such as pain, touch, temperature, pressure and vibration. Nerves in the skin also instruct skeletal muscles to shiver to produce body heat.

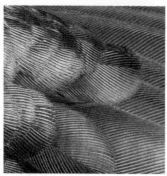

(Top) Feline fur.
(Middle) Amphibian scales.
(Bottom) Pennaceous bird feathers.
(Photographs courtesy of
I. Mazzoleni, D. McShaffrey,
L. Mazariegos)

Vertebrate groups such as mammals, amphibians, reptiles, fish, and birds have multiple layers of skin that function in ways similar to human skin even though it may not be immediately obvious. Many of these groups have various protuberances originating from the skin, called appendages, which work in conjunction with the skin. While humans have small amounts of hair on their bodies, most other mammals have fur that serves to thermoregulate, protect, sense, and communicate. Like human hair, fur is composed of keratin although it is usually made up of two layers — ground hairs and guard hairs. Ground hairs are the bottom layer of fur that tend to be thick and serve as insulation. Guard hairs, the top layer of fur, are usually longer and coarser than ground hairs. Guard hairs contain pigmentation, which can help in camouflage or in attracting mates through coloration patterns. This layer also serves as protection from elements such as rain, because it often has water-repellent properties. Fur further aids in thermoregulation through nerves in the skin that respond to cold and heat. The nerves activate muscles in the hair follicles that contract to pull the hair shafts erect, creating insulating air spaces. When the shafts are flattened, less air is trapped in the fur and the animal is able to release heat. In some cases animal fur is modified into hard, spiny protrusions that can serve as further protection from predators. Fur also delivers sensory messages to the body. Other appendages originating from the skin include scales and feathers, which are also make of keratin. Scales are hard plates that serve in protection from the elements, predators, and even prey. Snakes and lizards are the animals most associated with scales; however, a particular group of mammals — pangolins — has scales as well, though they are actually modified hairs. Feathers, on the other hand, are an appendage unique to birds; they are often considered the most complex appendage in vertebrates. There are two main types of feather that serve somewhat analogous functions as the fur of mammals. Pennaceous feathers have hooks and barbs that lock together to provide a solid, stiffened surface for flight. Birds coat their feathers with wax secreted from a specialized gland that functions in waterproofing, through the conditioning of feathers, and parasite resistance. Plumaceous feathers do not have hooks and barbs and are therefore fluffy, allowing air to be trapped and provide insulation.

The term "skin" is not commonly used for the outer coverings of invertebrates. For example, insects, arachnids, and crustaceans all have what is called an exoskeleton. This is a hard structure that

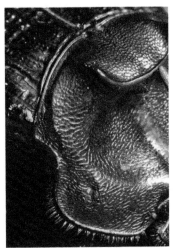

Dung beetle (*Scarabidae*) exoskeleton detail. (Photograph courtesy of S. McCann)

supports, covers and protects an animal's body. It can consist of a variety of compounds, but most often it is made of chitin. It is somewhat comparable to the function, but not structure, of keratin. Due to their rigidity, exoskeletons have a number of functions, including protection, sensation, a barrier to water loss, communication (through their coloration), and support. There is further variation in skin types that will be further explored in this book.

The Four Selected Functions To fully explore the relationship between animal skins and building envelopes, the book focuses on disparate animals, skin types, functions, and climates and attempts to demonstrate the variety of ways the developed methodology can be implemented and expanded by others. We have selected animals that show clear and, in many examples, unusual adaptations to four major functions of skin: communication, thermoregulation, water balance, and protection. These particular themes were selected because they are the most interesting and relevant to application in architecture and design. Thinking about how these functions are accomplished in nature can lead to innovative ways of providing human comfort, while lessening the built world's environmental impact through changing the ways in which we design.

Communication is crucial for animal survival, and strategies for communication take many forms. Modes of communication provide great insights to architectural investigations as designers learn from animals about the exchange of information within the built environment. We focus here on one form of communication widely used in nature: coloration. Coloration is used for warning, protection, camouflage, and sexual attraction. In architecture, some designers have explored the use of color as a communication tool, but more can be done by improving and diversifying the use of color through implementing the strategies of the selected organisms.

Thermoregulation is the ability most animals depend on to keep their body temperature within certain critical boundaries. Extremes in temperature pose tremendous physiological challenges to living organisms, and various mechanisms to help regulate temperature have evolved. Endotherms internally maintain their body temperature, whereas in ectotherms, temperature regulation is a function of their external environment. In either case, thermoregulation is achieved through remarkable physiological and behavioral processes. One of the major challenges architecture faces is to provide thermal comfort to human inhabitants, particularly in extreme circumstances, when human bodies cannot acclimate to external conditions. Animal systems have much to teach us regarding how to control temperature as well as how to limit energy expenditure in doing so.

Water balance is crucial for all organisms, considering that cells are composed primarily of water. Water is needed for numerous biochemical reactions, and it can dissolve and transport nutrients and other molecules. Animals have evolved many novel strategies to collect water and prevent water loss, particularly in water-limited habitats. In design, learning how animals prevent water loss may extend humans' ability to survive in dry and inhospitable conditions through implementing systems that minimize the use of water as well as collect, store and reuse it.

Protection from predators, parasites, physical injury, and the environment can occur in many ways. Humans can learn from the novel and complex adaptations that have evolved in the animal kingdom. Shelter, the most archetypical element in architecture, serves to protect or shield inhabitants from many things, and designers, by drawing inspiration from the efficiency of animal systems and the multiplicity of functions integrated within one system, can develop novel, responsive solutions.

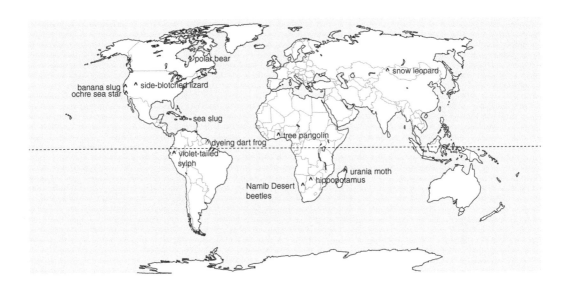

The animal species and localities chosen for the projects are shown on a world map. (Image courtesy of im studio mi/la)

3 Communication

The velvet-purple Coronet humming-bird has dazzling iridescent plumage used in courtship rituals and displays of territoriality. (Photograph courtesy of J. Rothmeyer)

Animal communication occurs when a signal sent by one individual has an effect on the behavior of at least one other individual. Communication happens via all senses and can occur through smell, touch, movement, gestures, sound, vibration, facial expressions, electrical signals, and coloration. These various forms of communication serve many functions: individual or species recognition, courtship rituals, warning off predators, aggression and territoriality, and signaling about food or other resources.

Animal coloration patterns can be powerful signals, because they aid in all aspects of communication. Colors are wavelengths of light; differences in wavelengths produce different colors. Bright and contrasting color patterns send the signal to predators that animals are poisonous or distasteful and therefore harmful to eat. Camouflage helps animals blend into their surroundings; prey can use camouflage to avoid predators, while predators use it to sneak up on prey. Certain coloration serves to confuse or dazzle predators. Many animals mimic the color patterns of other animals, usually to avoid being eaten. Colors also serve to promote sexual attraction and therefore reproduction.[11]

Animal coloration is due primarily to pigmentation. Pigments are chemicals produced inside tissues such as the skin, skin appendages, and even internal tissues and organs. There are many kinds of animal pigments; the most important in the color of animal coverings are melanins, carotenoids, and pterins.[12] Some specialized cells do not contain pigment but still function in coloration.[13]

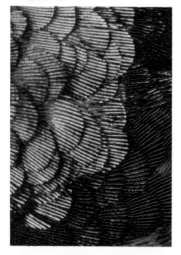

Coloration can also be due to microscopic changes in the physical structure of an animal's surface, or a combination of pigmentation and structure. The physical characteristics of the surface can cause the scattering, interference, and diffraction of light, leading to iridescence.[14]

Animals perceive color in many ways, and there is a great range of ability to detect different colors (wavelengths) among animals. Some animals can only detect the difference between light and dark, while others have sophisticated eyes that perceive many colors and sharp images, and have good depth perception. Many aquatic animals have evolved excellent color perception, because objects underwater are harder to distinguish based on contrast against the background. Bees, some birds and perhaps a few mammals can see UV, or polarized light, because objects in their environment absorb it. Most birds see in color and, in fact, have colored oil droplets in their eyes that help them have even better color vision than humans. Most mammals, however, have limited color vision; humans and other primates are the only mammals with a well-developed ability to perceive color.

Designers can find inspiration from understanding the relationship of an organism to its ecosystem as well as how humans perceive their color and how that perception can be useful in design. Most organisms do not see color as humans do, so communication between animals cannot be duplicated by humans. However, we are ultimately designing for the human perspective; that is the priority of the design process. The case studies presented embrace different strategies for coloration and are used for different architectural purposes.

Buildings communicate in a variety of ways, at times as status symbols and visible icons and at other times blending in invisibly and passing unobserved. There are four aspects of particular importance in design, two of which stem from modernism — extreme building height and façade transparency — and two of which are more contemporary and have been looked at in this book — color and camouflage.

Historically, façades have been important elements symbolizing power and strength, designed to communicate corporate identity and establish visual supremacy with their presence in the city. Skyscrapers, the equivalent of Middle Age towers, are the

(Top) The bright colors and iridescence of a hummingbird's feathers can serve to attract mates, to display aggression, or to warn off predators. (Photograph courtesy of L. Mazariegos)
(Bottom) Males and females of the cobalt blue tarantula have iridescent blue legs until they reach the final stage of sexual maturity, when, oddly, males lose their iridescence and become brownish in color. (Photograph courtesy of Flamesbane Wikimedia)

modern icons of power. Starting with modernism the extensive use of glass allowed these tall beacons to convey not only physical transparency but also a conceptual one, a manifestation of corporate operational transparency for public shareholders. Exemplars are Ludwig Mies van der Rohe's New York Seagram Building (1958) as well as the Chicago Sears Tower (1973) by Skidmore, Owings & Merrill (SOM).

Beyond the use of transparency in recent decades, few architects have ventured into using coloration as a means of communication. Each project of Berlin's Sauerbruch Hutton Architects uses color as the primary means to establish an architectural presence in the urban environment; the building massing is articulated with gradients of colors: greens turning into oranges turning into pinks, leading the viewer through a "rainbow" that dissolves the building presence from a single mass into a fragmentation of individual unique elements. To the contrary, Emilio Ambasz's projects often bury themselves under layers of earth, camouflaging and causing the building to disappear under strata of grass, hiding away while providing the benefit of thermal mass to the building's inhabitants. In addition, these buildings often provide the city with a green public space.

Nature can provide functional inspiration for novel uses of color in design. By studying the reflective microstructure of the Urania moth's wing scales, a project was designed that uses discarded colored plastic bottles as a signal to inform villagers when water is available. The iridescence of the violet-tailed sylph hummingbird inspired a colorful pavilion made by interlocking highly reflective panels to communicate about activities occurring in the pavilion. The photosynthetic lettuce sea slug inspired a center that uses different species of algae to produce biofuels while simultaneously providing a chromatically dynamic beacon.

Panel facade composed of timber and polychromatic glass. Sauerbruch Hutton Architects, The Federal Environment Agency, Dessau, 2005. (Photographs courtesy of Sauerbruch Hutton Architects, bitterbredt.de and Annette Kisling)

Animal Examples — Communication The black and white stripes of zebras can confuse predators that cannot see in color, and also make them unappealing to biting insect parasites. Leafhoppers and thorn insects use both coloration and appendages to make them difficult to distinguish from their environments. Some reptiles, such as the chameleon, are able to voluntarily regulate the size of pigment holding cells and to change color as they move from place to place. The chameleon also uses this ability to send signals to members of its own species. The bright coloration of the blue-banded goby appears to stand out to the human eye, but in vividly colored reefs it is hard to spot. The leafy sea dragon floats slowly through the water like a tangle of seaweed. The predatory stonefish is colored and ornamented to camouflage among the rocks on the sea floor as it lays in wait for passing prey.

Animals ranging from invertebrates to large mammals use coloration for a multitude of purposes. (Photographs courtesy of: zebra, J. Rothmeyer; blue-banded goby, M. Bartosek; chameleon, K. Tolley; stonefish, B. Larison; leafhopper, S. McCann; leafy sea dragon, I. Mazzoleni; thorn insects, M. Hedin)

Animal Examples — Communication Flamingoes are naturally white, but develop a pink coloration as carotenoids from their shrimp prey are incorporated into their feathers. Dragonflies use color to warn off predators. The bright, colorful plumage of the male mandarin duck is permanent and its intensity provides a measure of his genetic condition to the dull brown females. Both sexes of the scarlet macaw are boldly colored; this might facilitate recognition among members of the same species in dark rainforests, but scientists are still unsure. The dominant male mandrill in a group develops a brightly colored face that signals his status and restricts breeding rights to him alone. The male anole fans a brightly colored flap of skin called a dewlap from his throat to advertize his presence and condition to females in the area. In ladybugs, black spots stand out against a red background to warn predators that they are distasteful.

Animals ranging from invertebrates to large mammals use coloration for a multitude of purposes. (Photographs courtesy of: flamingo, morgueFile.com; dragonfly, B. Larison; male anole & ladybug, S. McCann; scarlet macaw, A. Kirschel; male mandril, Wikimedia; mandarin duck, M. Montese)

Animal Examples   Many insects possess striking coloration patterns; the function of these is likely to ward off predators or attract mates. The saddleback caterpillar is bright green with a white and brown circle resembling a saddle. They have irritating, venomous hairs on the protrusions on the front and back of their bodies, which prevent them from being eaten by predators. Cuckoo wasps have vibrantly colored metallic bodies, leading to the additional common names of jewel wasps or emerald wasps. Many beetles are iridescent, including the family called the metallic wood boring beetles. This family has often been used in human ornamentation and decoration because of their dazzling coloration. Glasswing butterflies have partially transparent wings. This may allow them to camouflage from predators by blending in to their surroundings. The physical mechanism that causes transparency is not known, but may be due to having fewer scales.

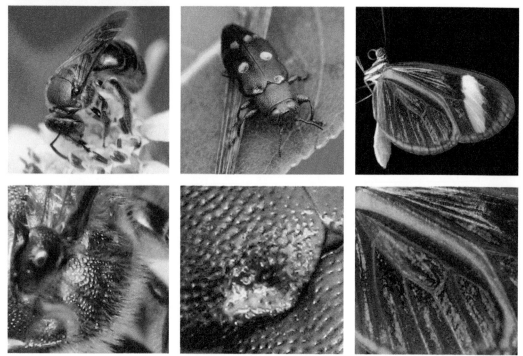

Insects possess a lot of variation in coloration, including having bright, iridescent or even transparent coloration. (Photographs courtesy of: saddleback caterpillar, cuckoo wasp and metallic wood boring beetles, S. McCann; glasswing butterfly, S. Yanoviak)

Urania moth
Chrysiridia rhipheus

Phylum:	Arthropoda	Family:	Uraniidae
Class:	Lepidoptera	Genus:	*Chrysiridia*
Order:	Pholidota	Species:	*C. rhipheus*

Photograph courtesy of A. Richards, Bohart Museum of Entomology, University of California at Davis.

Habitat & Climate   The Madagascan sunset moth only occurs on the island of Madagascar. The island's climate varies considerably longitudinally. The eastern part of the country supports a rainforest habitat due to trade and monsoon winds along the east coast causing high precipitation levels. The winds lose moisture while moving west, so the central part of the island is drier; it is also cooler due to the higher altitudes. The western part of the island is mostly arid, with semi-desert conditions in the southwest. Two principal seasons are defined: a hot, rainy season occurring from November through April and a cooler, drier season occurring from May through October. The sunset moth is present throughout most parts of Madagascar, but they migrate across the country throughout the year in response to changes of the host plants they depend on for survival in the early stages of life.

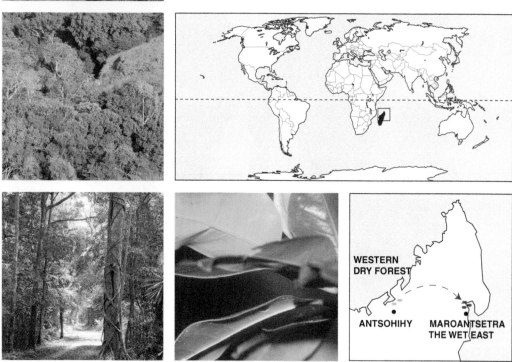

Dry deciduous forest in the western region of Madagascar. Close up of *Omphalea's* leaves, the moth's host plant. (Photographs courtesy of A.P. Raselimanana)

Animal Physiological, Behavioral & Anatomical Elements
The behavior of *C. rhipheus* is atypical of many moths, because it is active during the day. The coloration patterns on the moth are warning signals to let potential predators know the moths are toxic. The toxicity comes from trees and shrubs in the genus *Omphalea*. Caterpillars feed on the toxic plant components, which are stored through their development, even in to the adult stages.

The dependence on the host plant is so strong that sunset moths migrate to different populations of their host plant during the year; specifically, they migrate from species in the dry forest in the west to the rainforest in the east. The moth caterpillars negatively affect the *Omphalea* populations, because they eat the flowers and fruit from the entire plant.

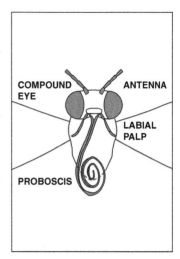

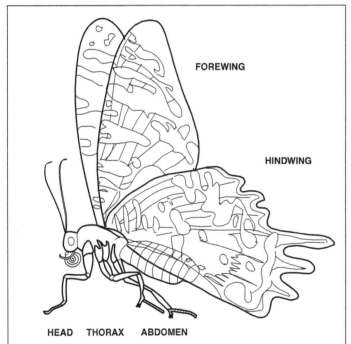

C. ripheus caterpillars feed exclusively on *Omphalea* plants, often completely defoliating them. They eat the leaves, flowers, fruit, tendrils and new shoots, causing damage to all parts of the host plants. (Photograph, Wikimedia; project team: B. Frati & N. Zarfaty)

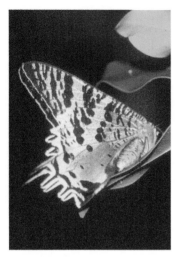

Interface between the Skin & External World The most striking feature of the Madagascan sunset moth is its multi-colored wing patterning, including areas of brilliant iridescence. Iridescence is defined as the surface of a structure that changes color when the viewing angle or angle of lighting is changed. Iridescent colors are structural colors; they tend to be brighter and purer than colors produced by pigments. Butterfly and moth wings are comprised of thousands of partly overlapping scales, and the structure of the scales affects the way they reflect light. Sunset moths have two layers of scales on their wings: the inner layer and cover scales. Each cover scale is a multilayered structure made up of thin layers of cuticle between thin layers of air. The layers of cuticle are supported by very small, spaced out columns of cuticle. The layers of cuticle and air can be different thicknesses, which create the variation in color in the wings.

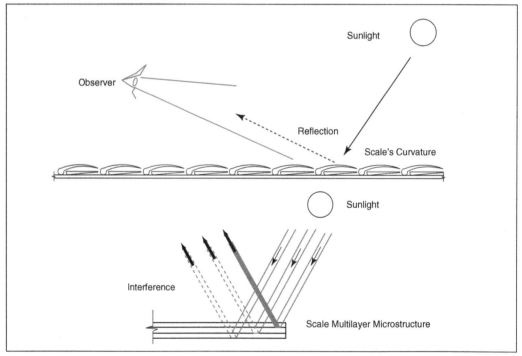

In iridescence, which is caused by structural modifications, the hue of the color changes with the angle of the observer and light shining on the surface. The brilliant colors are caused by the reflection and interference of light going through multiple surfaces. (Photograph courtesy of A.P. Raselimanana)

Interrelationship between the Skin & Internal Systems
Different pigments can occur within scales, and with the combination of the multilayered cuticle surface, different iridescent colors can be obtained. When light shines through the first layer of cuticle, some is reflected back from the surface and some goes through to the next level. With each level the light penetrates, more light is reflected from the next level down. The reflected light from each level travels the same way, that is, away from the cuticle layers.

Light occurs in wavelengths, and if the crests of the wavelengths overlap with each other, they amplify the effect of reflection. When the wavelengths line up with each other intense coloration occurs. This phenomenon is called constructive interference, and it is what causes iridescence.

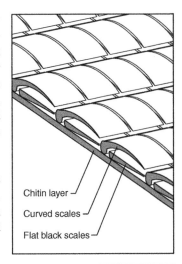

Chitin layer ⏋
Curved scales ⏋
Flat black scales ⏌

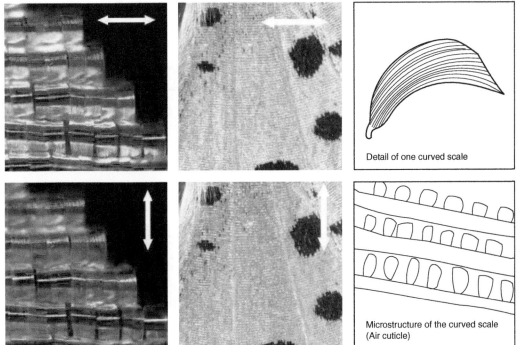

Detail of one curved scale

Microstructure of the curved scale
(Air cuticle)

In the panels the white arrows represent the direction of light shining on the scales and illustrate how wing coloration is a function of angle of perception. The wings of the moth are made of two layers of scales: an upper, curved layer and a lower black layer of flat scales. Both layers of scales are attached to a chitin layer. The microstructure of each curved scale is responsible for producing the brilliant iridescent colors in the sunset moth. (Micrographs courtesy of Yoshioka and Kinoshita [2007])

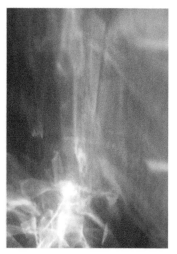

Proto-Architectural Project The project is located in the moth's natural habitat, an arid town in western Madagascar. The design for the building envelope was inspired by the reflective and refractive properties of the moth's multilayer and multi-color wing assembly. The building provides a new system of clean drinking water for the local community where rainwater is collected through a large roof surface. A filtration system which forces water from the roof into the building's envelope is constructed to support the aggregation of reusable colored bottles that collect water through a structural trellis. The angle of the bottles changes depending upon the amount of water collected in the bottles: when empty the bottles are perpendicular to the façade and as the container becomes heavy with water it shifts parallel to the façade. Light plays with the shape and angle of the water-filled bottles, creating spectacular moments of reflection.

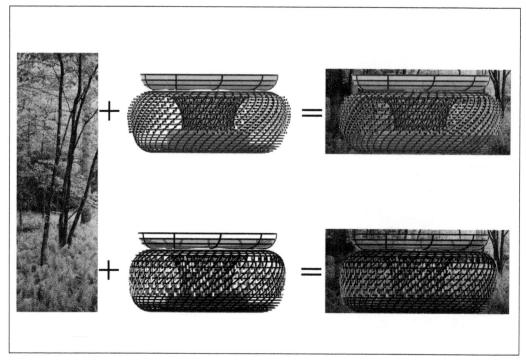

The coloration effect provided by the plastic bottles was analyzed through a series of experiments investigating the relationships between a surface, material, water, light and the resultant colors.

Project Documentation   The façade communicates water availability by displaying different parts of the colored bottles. The base of the bottle is black, while the remaining part is colorful, either blue or green, typical colors used to enhance passive solar purification processes.

The bottles' layer is sequenced to take advantage of the play of light through the bottles' curvature to create varying moments of reflection. This layering technique, together with the overall circular geometry of the structure, alters the angle of reflection as each bottle has a unique position in relationship to the sun, thus providing a kaleidoscopic effect. The façade explores both the idea of color as a useful means of communication and the physical properties of light and color.

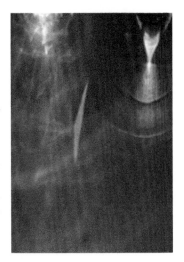

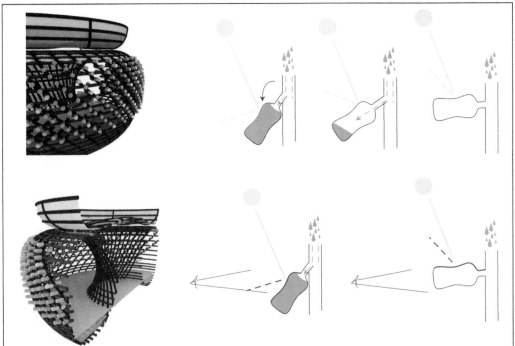

Color is used to attract people to the building and is created through the superimposition of different façade layers, with the light passing through the changing angles of the water-filled multi-colored bottles.

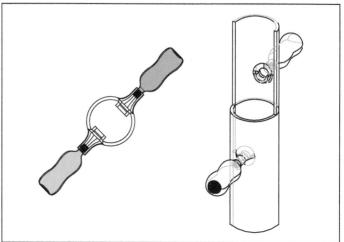

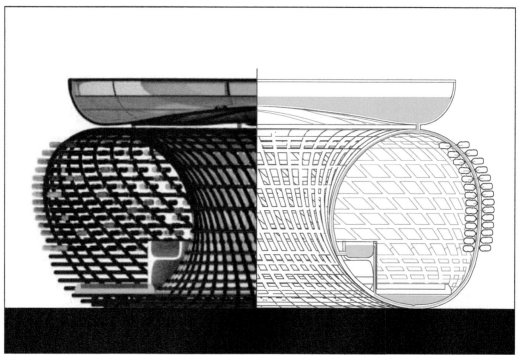

The curved surface of the tubular structure and the variable distance between tubes is responsible for reflecting and refracting color, which varies through time to announce water availability. The project proposes to reuse locally collected material such as plastic tubes, pipes and bottles, an environmentally conscious and low budget strategy.

Water scarcity and pollution are dire problems in developing countries. Finding simple ways to collect and purify rain water is crucial to the well-being of these communities. Additionally, re-purposing the uses of plastic bottles, by extending their life cycle, minimizes landfill and illegal toxic burnings.

Animal Examples Communication is not always delivered using aural mechanisms and many organisms communicate via visual display. Birds in particular have mastered visual communication. Bright hues and iridescent structures can act as a signal of quality, territory defense, or even as a distraction to predators. The violet eared hummingbird uses purple and green iridescence created by alternating layers of refracting light striking pigments to attract females. The resplendent quetzal flashes bright blue, a color created by light reflecting off trapped water molecules within the feather. The fiery throated hummingbird combines these cooler colors with bright reds and oranges created by physical pigments in the feathers. The swallow-tailed bee eater combines these feather colors with a striking red eye pigment to capture an observer's attention, be it a fellow bee eater or a human.

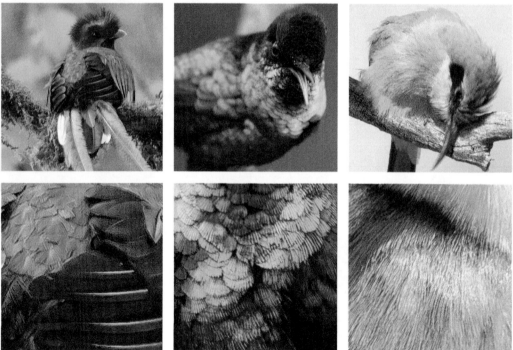

The variation of colors in bird feathers acts as a form of communication vital to attracting the attention of the opposite sex, as a warning to competitiors, or even as a camouflage in a dense forest. (Photographs courtesy of violet-eared hummingbird, resplendent quetzal, fiery throated hummingbird and swallow-tailed bee eater, J. Rothmeyer)

Violet-tailed sylph
Aglaiocerus coelestis

Phylum:	Chordata	**Family:**	Trochilidae
Class:	Aves	**Genus:**	*Aglaiocercus*
Order:	Apodiformes	**Species:**	*A. coelestis*

Photograph courtesy of J. Rothmeyer

Habitat & Climate   The violet-tailed sylph (*Aglaiocercus coelestis*) can be found along the western slopes of the Andes, but is most abundant in the cloud forests of Ecuador. These types of forests are moist, due to the high level of precipitation that falls yearly; the rainy season from January to May is cool, 8–15°C (46.4–59°F), due to the high altitude, from 500 to 4000 m (~1640–13,123 ft). This combination of moisture, 50 to 1000 cm/year (19.5–393.5 in), and elevation creates slow evaporation, resulting in a lush, densely forested habitat. The eastern and western slopes of the Andes in Ecuador and Colombia experience rainfall fed by moist air coming from the Amazon Basin. Precipitation is high due to winds that bring in water vapor from the Amazon. Such a unique climate allows for a variety of exotic plants to grow and flourish. The *Heliconia* is one such plant, which produces nectar that the violet-tailed sylph feeds on.

The violet-tailed sylph has a home range that includes forest edges, shrubs, and deep cloud forests within Ecuador and Colombia. This species is unique in its altitudinal range and is most easily spotted around 1000 meters above sea level. These cloud forests are unique within the Andes and harbor much biodiversity and endemic species. (Photographs courtesy of J. Rothmeyer)

Animal Physiological, Behavioral & Anatomical Elements A hummingbird's normal body temperature runs around 40.5°C (105°F). When a hummingbird sleeps, this temperature will drop to as low as 21°C (70°F). Due to its small size, the hummingbird rapidly loses heat to the outside world because of the ratio between its volume (small) as compared to its relative surface area (large). This fact, combined with the high, cool elevations where these hummingbirds often reside, makes heat conservation a priority. To conserve energy a hummingbird will go through a nightly rapid transition in metabolism and activity equivalent to a brief hibernation. This state is called torpor and is so extreme that one can pluck a hummingbird off a branch while in torpor without the bird even moving! Always on the search for energy sources, many hummingbirds are nectavores and seek out this high-energy food source even in the most protected of flowers.

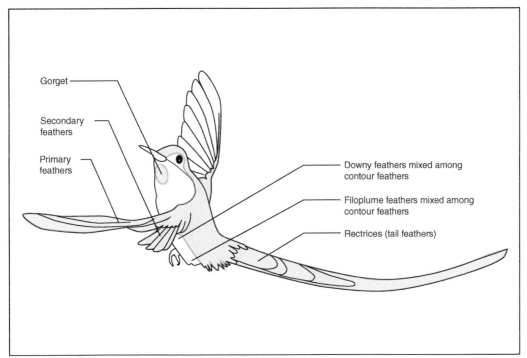

The most brilliant iridescent colors are found in males and have two purposes: to advertise their physiological quality to other males and to attract females. Feathers carry sub-microscopic structures that produce bright colors; however, the brilliance can be seen only with certain positions involving the bird, sun, and observer. (Photograph courtesy of J. Rothmeyer; project team: J.M. Helinurm & A. Amiri)

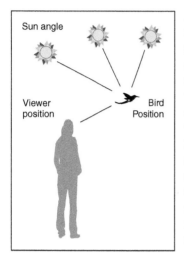

Interface between the Skin & External World Hummingbirds' feathers are functionally similar to other birds, yet produce unique iridescent colors. The shaft of the feather (or rachis) extends out as alternating thread-like barbs and barbules, which grow out at proper angles in order to hook with one another and maintain structural integrity, but at the same time remaining flexible enough to bend or disconnect without breakage. Feathers lie on top of each other, keeping only the iridescent tips visible. The keratin protein found in feathers provides flexibility and integrity, and acts as a natural insulator. Contour feathers cover the body and appendages of birds. These feathers have an expanded vane that provides the continuous surface necessary for flight. Changes in the angles of these feathers allow for changes in the amount of feather brilliance observed from different positions. Some feathers are flat; others are curved.

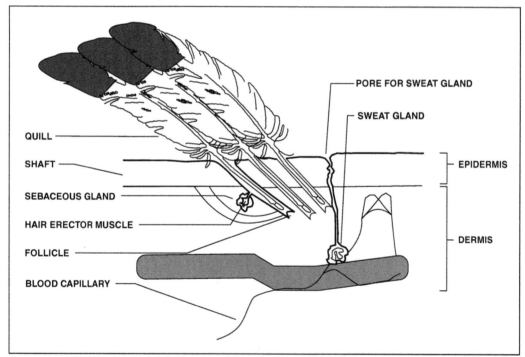

Some feathers, for instance on the gorget and crown, resemble flat mirrors, and light that hits them can be reflected in only one direction. The sun, observer (bird or human), and bird must be aligned properly to view the brilliance of the plumage. When no light shines on these feathers they appear black. Other feathers, like those on the back, are curved inward to resemble concave mirrors. These curves come in many angles, allowing for light to be refracted to the observer at different directions. The section shows feathers insert into bird skin through the epidermis and dermis.

Interrelationship between the Skin & Internal Systems To be as lightweight as possible, most of the hummingbird's bones are extremely porous, and their legs are extremely small, short, and stubby. Perhaps the most diagnostic character of the hummingbird is its wing structure, which enables it to hover, fly backwards, and to change direction with remarkable precision. Several hummingbirds can even fly upside-down for a short period of time. The shoulder joint is a ball and socket joint that allows the hummingbird to rotate their wings 180° in all directions. When hummingbirds take flight, they move their wings in an oval pattern and maintain their body in an upright fusiform position, with their entire body facing the world. When they are hovering they move their wings in a figure-eight motion. A hummingbird can fly at an average speed of 40 to 48 km/h (~25 to 30 mph) and dive at a speed of up to 96.5 km/h (60 mph).

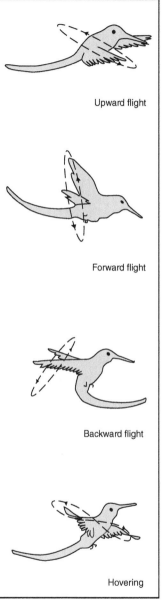

Upward flight

Forward flight

Backward flight

Hovering

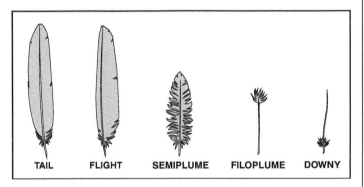

TAIL   FLIGHT   SEMIPLUME   FILOPLUME   DOWNY

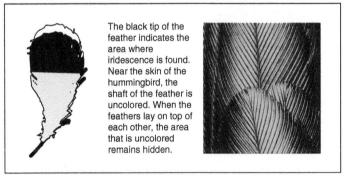

The black tip of the feather indicates the area where iridescence is found. Near the skin of the hummingbird, the shaft of the feather is uncolored. When the feathers lay on top of each other, the area that is uncolored remains hidden.

Hummingbird wings beat about 70 times per second while in regular flight, and up to 200 times per second when diving. To power this movement, their metabolism is extremely fast, and heart rates average 250 beats a minute, with a maximum heart rate recorded of 1260 beats a minute. (Photograph courtesy of L. Mazariegos)

Proto-Architectural Project   The pavilion is located in Parque la Panecillio in Quito, Ecuador, a city 2800 m (9185 ft) above sea level and 24 km (15 mi) south of the equator. Taking advantage of the subtropical, yet cool climate (due to its elevation), the pavilion develops a porous envelope, enabling the elements to permeate through. Inspired by the iridescence of the feathers, the pavilion takes advantage of the color palette found on the violet-tailed sylph: blues, violets, and greens. Color operates as a device for aesthetics and a significant means of communication. It serves as an indicator for occurring and upcoming events by radiating colors in response to dynamic atmospheric changes. Bright colors are a device to express energy and enthusiasm, thus enhancing the sensorial and spatial experience of the inhabitant – from within and afar.

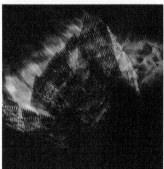

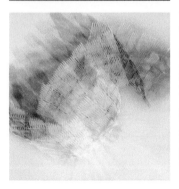

Iridescence is dependent on the position of the object, the observer and the light rays hitting the object. Inspired by these factors, 3D digital modeling techniques were utilized to explore the articulation of individual panels in relation to their rotational array – with the intent to enhance and maximize the reflectivity of sunlight, while minimizing the quantity of artificial light sources at night. (Photograph courtesy of L. Mazariegos)

Project Documentation   The pavilion's panelization and joint system is inspired by the hooked barbules in a hummingbird's feather and the implementation of principles of light refraction magnified by water. One level of refraction is developed through the panels' materiality, composed of two layers of semitransparent resin and a thin layer of glass to provide additional qualities. A second level of refraction is established by understanding how light reacts differently when it hits and travels through water – resulting in variations in the colors of light dispersed. The hook joint creates small water pockets between the panels, thus enhancing the desired light deflection. Resource optimization and efficiency are crucial to the pavilion and its use as public space, hence the exploration and strategic implementation of a repeated module to produce communicative effects.

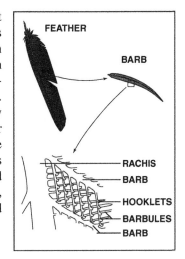

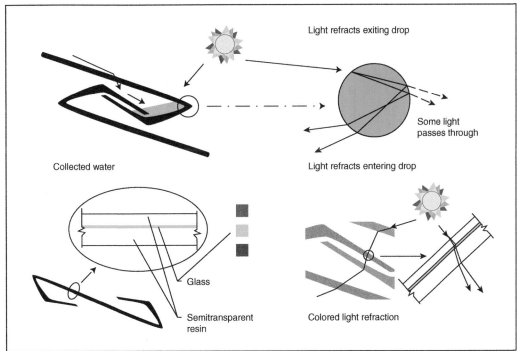

Mimicking the complex structure of feathers, the modules overlap, thereby magnifying the light refraction between layers, each made of glass and colored resins.

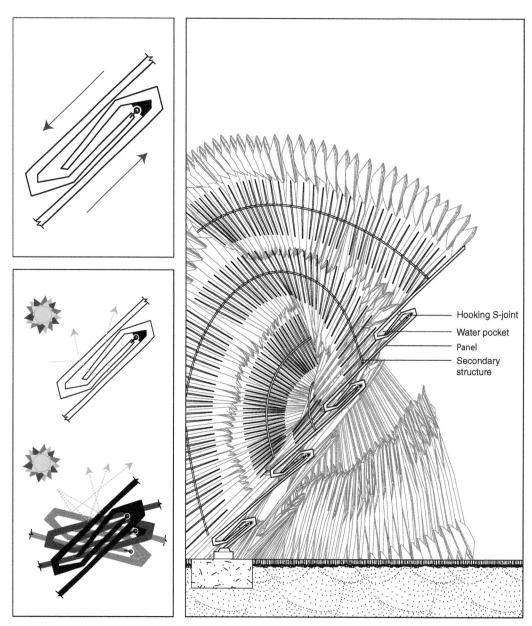

Hooking S-joint
Water pocket
Panel
Secondary structure

The structure displays a broad spectrum of color variation throughout the pavilion. The modular panels are arrayed to reflect and refract the sunlight at varying angles. The color brilliance is ultimately enhanced through the combined effects of the semitransparent panels and water pockets. The panel joints form an S-like shape to interlock with one another, forming a small cavity to trap water.

The pavilion acts as a beacon during activities and events, becoming a glimmer of coloration when observed from the city. The colorful variations of refracted light infiltrate the internal spaces, amplifying the sensorial experience promoted by the physical environment.

Animal Examples   Marine invertebrates use coloration for different purposes. Giant clams are vibrantly colored, often with iridescent patches on their mantles, or body walls. They achieve this coloration through symbiotic algae harbored in their tissues. The algae are given a protected place to live, and in turn, the clam ingests the sugars and proteins produced by the colorful algae. A group of sea slugs, called nudibranchs, are known for their vivid coloration.

Color patterns, often contrasting, are thought to serve as warning coloration to notify predators of their distastefulness or toxicity. The blue-ringed octopus is extremely venomous. They are small and normally camouflage with their surroundings. However, when agitated, their skin turns bright yellow, and the blue and black rings covering their bodies darken substantially when threatened.

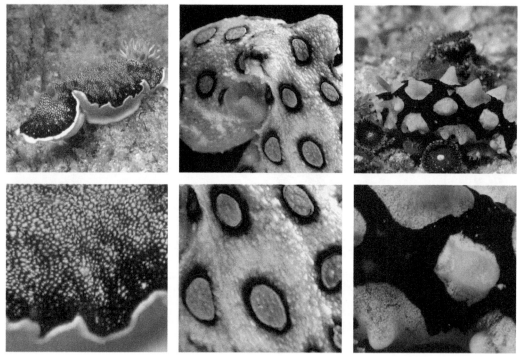

Vivid colors are thought to serve as warning coloration to notify predators of their distastefulness or toxicity. Color patterns are often contrasting, like red paired with white or black paired with yellow, to be particularly conspicuous to predators. On the other hand, coloration can serve as a camouflaging mechanism through which certain animals can establish symbiotic relationships with others. (Photographs courtesy of: giant clam, red sea slug (*Chromodoris* sp.) and blue sea slug (*Phyllidia* sp.), B. Larison; blue-ringed octopus, J. Himes)

Lettuce sea slug
Elysia crispata

Phylum:	Mollusca	**Family:**	Plakobranchoidea
Class:	Gastropoda	**Genus:**	*Elysia*
Order:	Heterobranchia	**Species:**	*E. crispata*

Photograph courtesy of P. Krug

Habitat & Climate The sea slug *Elysia crispata,* often referred to by aquarists as the lettuce slug, can be found on island coasts throughout the Caribbean Sea. Its habitat is limited to the shallow waters of coral reefs and mangrove lagoons, approximately 1.5 to 12 m (5 to 40 ft) in depth. They live in areas where water currents are relatively weak and temperatures range from 20–30°C (68 to 86°F) . Water temperatures closely mimic air temperatures in the Caribbean, with little change across seasons during a given year; that is typical for tropical regions due to their close proximity to the equator. Air temperatures are warm (20–32°C, 68–86°F,) and relative humidity is typically high (50–90%). Slight seasonal shifts do lead to rainy seasons in the tropics. In Haiti, the specific island chosen for this study, this increased rainfall occurs primarily from May to July.

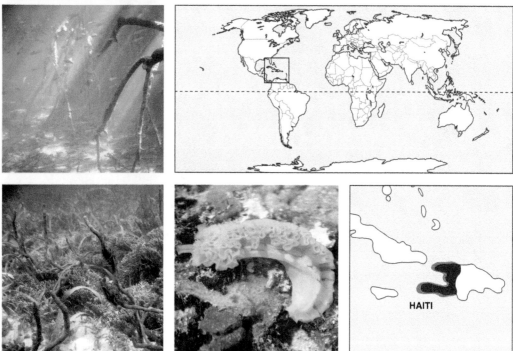

Elysia crispata lives on coral reefs and in mangrove lagoons of the Caribbean Sea, where it eats green algae and can store the algal chloroplasts in its own tissues. It is often found "basking" in the sun in order to provide energy for photosynthesis, making it solar powered. (Photographs courtesy of R. Ellingson)

Animal Physiological, Behavioral & Anatomical Elements   A unique characteristic of *Elysia crispata* is its ability to take functional chloroplasts (plastids) from its green algal food source and store them, a process called kleptoplasty. The slugs can be found "basking" in clear and shallow waters, absorbing the sun's energy to drive photosynthesis (the production of energy from light, water and carbon dioxide). This allows *E. crispata* and some of its relatives to go weeks or even months without food, making them the only known solar-powered animals. The common name "lettuce slug" comes from the leafy appearance of its parapodia, two large flaps that run the length of its body. Basking refers to the opening of these parapodia to absorb light, made more efficient by the increased surface area of their ruffled, leafy form. Stored plastids cause *E. crispata* to usually appear green, but vibrant shades of blue, yellow and hints of red can often be seen as well.

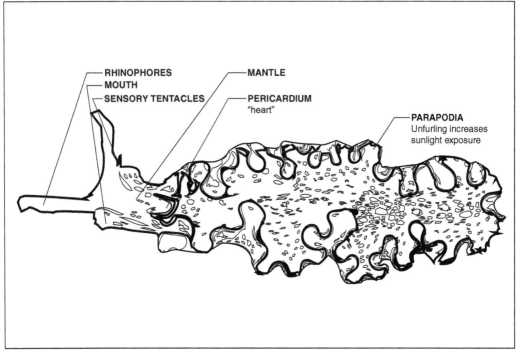

RHINOPHORES
MOUTH
SENSORY TENTACLES

MANTLE
PERICARDIUM
"heart"

PARAPODIA
Unfurling increases
sunlight exposure

E. crispata has a translucent body with a variation of colors. The rhinophores protruding from the back can range from blue to red to green. The slug's color is known to fade or intensify with the sequestration of chloroplasts within its tissues. The more recently a slug has eaten, the more colorful it will appear. *E. crispata* has the ability to expand its parapodia in order to capture more energy from the sun, or fold them in close to decrease its vulnerability. (Photograph courtesy of P. Krug; project team: A. Munoz & R. Hopkins)

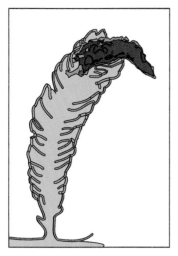

Interface between the Skin & External World The skin of *Elysia crispata* is translucent. Visible colors therefore come from internal tissues and organs or any material that might be incorporated into the skin. Translucent skin is particularly important for kleptoplasty since radiant energy from sunlight must reach internally stored chloroplasts. The leafy appearance of parapodia is crucial for increasing the surface area of the skin, allowing for efficient uptake of light for photosynthesis. The process of kleptoplasty causes the translucent skin to change in color from clear or pale to a rich green (or even blue), much like the leaf of a plant. Sea slugs have no shells to provide physical protection from predators. Therefore they rely on cryptic coloration, often blending in to the green color of their algae food, as well as large amounts of mucus produced by the skin. This mucus is distasteful to most predators and can often be toxic as well.

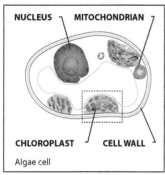

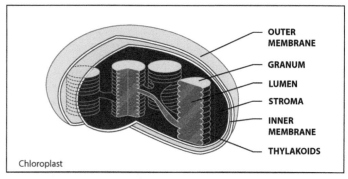

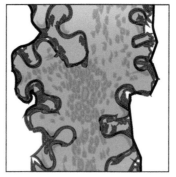

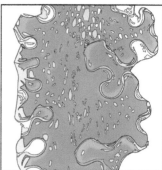

The chloroplast, contained in the algae cells, is the organelle responsible for photosynthesis. The slug has a slimy skin that provides no physical protection. Instead, it produces mucus that deters predators because of its bad taste and toxicity. Because they often adopt the green color of their algal food, camouflage may also be a useful tool to avoid predation.

Interrelationship between the Skin & Internal Systems   Factors that determine the color of *Elysia crispata* are tightly linked to the system it uses to harness energy. Since functional chloroplasts are sequestered in its skin, *E. crispata* usually appear green. Kleptoplasty causes the slugs to be green in much the same way that chloroplasts turn the leaves of plants green. Much of the light at wavelengths near the red and blue ends of the visible light spectrum are absorbed and used for photosynthesis, while most of the green light is reflected back to be observed from the outside. The digestive tract of *E. crispata* runs throughout the body and its vein-like patterning is often apparent, especially after a meal. Chloroplasts are stored in close proximity to the digestive tract to aid in the absorption of carbohydrates produced by photosynthesis. If *E. crispata* goes an extended period of time without food, few chloroplasts are present and its color becomes very pale.

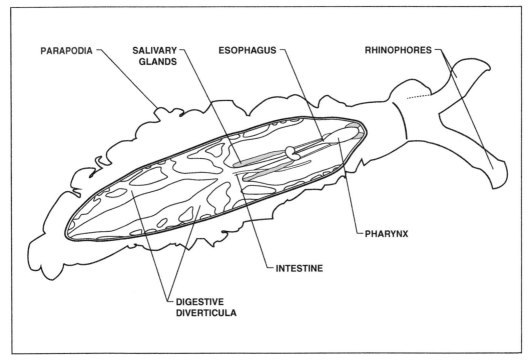

Above: Close up view of *E. crispata* showing individual chloroplasts (tiny green spots) sequestered in its tissues. Below: Illustration of parts of the lettuce slug's internal anatomy. (Photograph courtesy of P. Krug; diagram redrawn from R. Ellington)

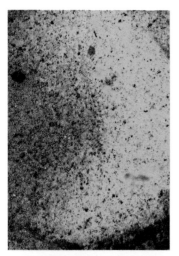

Proto-Architectural Project Based on the eastern coast of Haiti, the building intends to fill the need for a dependable energy source during the time of reconstruction after the 2010 earthquake disaster. The envelope system, inspired by the processes that generate color in *E. crispata*, is a photo-bioreactor made of layered plastic modules and tubes filled with algae.

Using locally abundant resources — ocean water, sun, and micro-organisms — the building produces biofuel, enabling it to change colors, ranging from red to blue to green, as the algae matures. The growth and harvest of the algae provide a dynamic beautification of the structure that communicates its level of energy production.

The building's coloration is obtained with the use of living organisms such as algae. By communicating the harnessing of energy through color change the enclosure produces a presence along the coast of the island.

Project Documentation Different algal species grow under unique conditions; therefore, a diversification of the species used allows optimization of biofuel production. As a by-product, variation in color display is also achieved. A system of interlocking tubes brings water to the plastic modules, which are layered depending on particular sun exposure requirements. At any given time there can be as many as six different species of algae maturing to create biodiesel, each exhibiting a different color. The weather and seasons play an integral part, as the sun dictates how fast or slow the algal growth process occurs.

Additionally, tubes filled with one more bioluminescent species of algae are intertwined to the structural elements, thus providing a beacon at night.

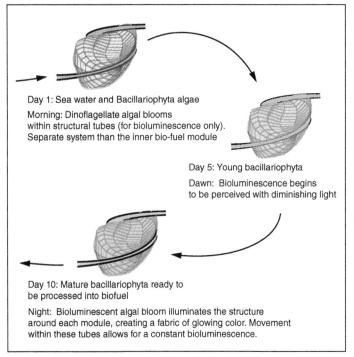

Day 1: Sea water and Bacillariophyta algae

Morning: Dinoflagellate algal blooms within structural tubes (for bioluminescence only). Separate system than the inner bio-fuel module

Day 5: Young bacillariophyta

Dawn: Bioluminescence begins to be perceived with diminishing light

Day 10: Mature bacillariophyta ready to be processed into biofuel

Night: Bioluminescent algal bloom illuminates the structure around each module, creating a fabric of glowing color. Movement within these tubes allows for a constant bioluminescence.

Algae are living organisms that use sun, energy, and CO_2 to photosynthesize. Mimicking the translucent affect of the *E. crispata* skin, the building's hollow tubular structure of the envelope contains bioluminescent algae, creating a light source during the night. The diagram indicates a timeline of harvesting cycles. Images show day, sunset, and nighttime building coloration scenarios.

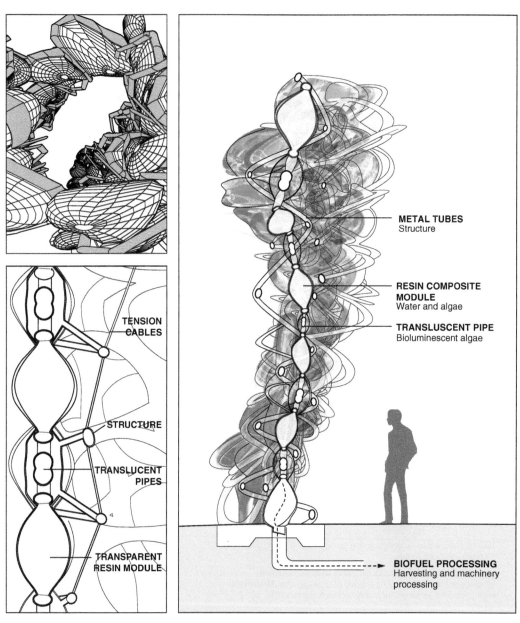

METAL TUBES
Structure

RESIN COMPOSITE
MODULE
Water and algae

TRANSLUSCENT PIPE
Bioluminescent algae

TENSION
CABLES

STRUCTURE

TRANSLUCENT
PIPES

TRANSPARENT
RESIN MODULE

BIOFUEL PROCESSING
Harvesting and machinery
processing

Networks of structural tubes support a series of interconnected modules which contain micro algae whose main purpose is to create biofuel. The translucent pipes contain bioluminescent algae whose main purpose is to provide bright coloration during the day and bioluminescence at night.

The multiple layers of sacs enclosing micro algae and the tubular structure containing bioluminescent algae create an array of ever changing coloration inside and outside the structure. The variation in color is produced both by the different types of algae and by their different level of maturation, providing a dynamic and continuously varying message of productivity.

4 Thermoregulation

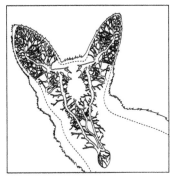

The large ears of the fennec fox are endowed with a network of blood vessels that can be used to flush excess body heat. (Image courtesy of D. Kim & J. Cambron)

Maintaining an appropriate body temperature is critical for all aspects of biochemical and physiological function, and animals expend a great deal of energy controlling their body temperature. There are two main strategies by which animals regulate their body temperatures. Ectotherms, which include invertebrates, fish, and reptiles, obtain most of their body heat directly from the environment, and their metabolic rates fluctuate with ambient temperature. Endotherms, such as mammals and birds, produce their own heat through metabolic processes and are able to maintain constant body temperatures. Almost all heat produced is ultimately lost, making endothermy an energetically expensive process. Consequently, endotherms need to eat high-energy foods or feed very often.

Both endotherms and ectotherms use behavioral strategies to manipulate their body temperature when the environmental temperature is either too hot or too cold. Heat is lost and gained through four major mechanisms: conduction, convection, radiation and evaporation. Conduction occurs when two objects of different temperatures come into contact with each other. For example, to gain heat an animal might lie on a rock that has been warmed by the sun; to cool down an animal might wallow in cool mud. Convection is similar but occurs when heat is transferred through air or water. This often occurs when an animal is hot and its body heat warms the air around it. As air passes over the skin, the mass of hot air is pushed away from the body and allows a cooler mass of air to take its place. Animals can seek out cold

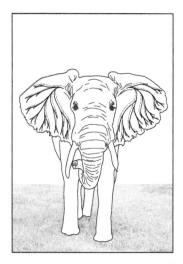

Due to their large body size and small surface to volume ratio, elephants retain a lot of body heat. Therefore, they flap their ears, bathe in water, and stand in shade to cool down. (Image courtesy of T. Barsegyan)

air or water currents to cool down in this way, and conversely, find warm air or water to warm their body temperature. When heat transfer takes place without physical contact this is called radiation. Objects give off heat in the form of electromagnetic radiation when their temperature is warmer than that of the surrounding environment. This means that animals can lose heat when their bodies are warmer than the ambient temperature, and they can gain heat from the sun when they are cooler. Finding shade to cool down or lying in the sun to warm up (basking) are ways of behaviorally thermoregulating using radiation. The only way animals can lose heat when the air temperature is warmer than their body temperature is through evaporation. Evaporation is a cooling mechanism that occurs when heat is released from the conversion of liquid water to water vapor. Animals cool themselves using evaporation through sweating and panting.

An animal's size, or more specifically, the ratio of its surface area to volume, is very important when considering heat gain and loss. As geometrically similar shapes get bigger, their surface area relative to their volume decreases. Because the surface is where heat is gained and lost to the environment, larger animals are therefore better at retaining body heat. This body mass to surface area relationship is why bigger animals tend to occur in cold environments and smaller animals tend to occur in hot environments. When large animals occur in hot environments, they tend to have less fur to facilitate heat loss. In addition to varying body size, evolution has acted to change the surface area of animals through creating longer and bigger extremities in hot environments and decreasing extremity length in cold environments. In hot climates longer extremities, such as longer limbs and big ears, allow more heat to be lost to the air, and shorter extremities can help prevent heat loss, thus conserving heat.

Animals that live in extreme heat and aridity have evolved many strategies to prevent overheating and to dissipate any excess body heat. These range from the physical attributes of the animals to physiological adaptations and behavioral changes. Many animals in hot climates are light colored to reflect heat. Long limbs not only allow for increased surface area but reduce radiation by keeping the body farther from the hot ground. Animals that cannot sweat may rub saliva over their bodies or urinate on themselves to promote heat loss via evaporation. Some animals have many blood vessels near the surface of their skin to promote convective

In cold environments ducks behaviorally thermoregulate by fluffing their feathers, tucking one leg under their breast feathers, and placing their bills in feathers. These strategies prevent heat loss. (Photograph courtesy of A. Holmberg)

heat loss. These vessels can dilate, or open further, to allow for increased blood flow and more heat to be dissipated. Some animals allow their body temperatures to rise with the environmental temperature during the day and then allow the stored heat to dissipate in the cool of the evening air. Many desert animals are only active at night or at dawn and dusk and stay in protected, shaded areas during the day. Small animals, particularly rodents, lizards and insects, may take refuge in underground burrows to prevent overheating in the hottest part of the day. Some rodents, frogs and arthropods are able to save energy by going into a kind of dormancy called estivation during the hottest months of the summer.

In contrast to the hottest climates, many animals are adapted to live in arctic environments where temperatures are extremely cold. Most arctic animals have some type of thick insulation in the form of fur, feathers, or blubber, a layer of subcutaneous fat. Fur and feathers can be adjusted to trap pockets of warm air close to the skin or release it when conditions are warm. Animals also curl up when at rest to minimize heat loss through reduction of surface area exposed to the environment. Under the skin, blubber prevents heat from being lost from the body core to the environment. Some

Lizards behaviorally thermoregulate by basking to acquire heat or retreating into shade to prevent heat gain. (Photograph courtesy of S. Yeliseev)

birds and mammals have a complex counter-current heat exchange system that allows warm blood coming from the heart to heat cooler blood coming from the extremities. This system may also occur in hot environments to cool blood flowing to temperature-sensitive body organs like the brain. Many ectotherms such as fish and insects survive subzero temperatures by producing proteins that prevent ice crystals from forming in their bodies. Additionally, endotherms such as mammals and birds may dig dens or huddle in groups to protect themselves from cold winds and temperatures. Just as desert animals reduce their metabolic rate in estivation, cold-adapted animals do the same thing when hibernating. This strategy minimizes heat loss by decreasing the difference in temperature between the animal and the environment.

Building envelopes are responsible for the mediations between indoor environmental comfort for the inhabitants and exterior climate conditions. In order to minimize the use of natural resources several passive strategies can be adopted. The most important of all is the placement of a building on a site and its orientation to environmental forces which can facilitate the use of rain, sun, and wind to achieve internal comfort.

For most of human history - and in a few populations today - nomadism characterized our way of life. This transient lifestyle was based on the need to find food which necessitated periodic

The envelope made by ethylene tetrafluoroethylene (ETFE) panels is activated using pneumatic mechanisms triggered by weather sensors. Cloud 9, Media-TIC, Barcelona, Spain, 2011. (Photograph courtesy of A. Suner)

movement to adapt to the changing seasons and food sources. Consequently, in nomadic societies shelters were not immobile; they had to be portable or relatively easy to construct. These kinds of structures, in comparison to the fixed ones most prevalent today, were less impactful on the environment. The rise of agriculturally-based societies led to a way of life in which human dwellings are designed for permanency, and thus the need to deal with the year-round weather conditions of a location.

For example, in desert regions thick walls provide thermal mass to help offset the high temperatures of the day while releasing heat during the cool nights. Conversely, in hot and humid climates, such as the tropics, raised, light structures allow for maximum cross ventilation, optimizing evaporation.

The excessive use of active, energy intensive systems places an undue burden on the planet. Passive strategies offer a valuable way to reduce our ecological footprint. This will help us achieve energy and carbon efficient buildings, more respectful of the environment and less demanding on the use of the planet' finite resources. Moreover, technology plays an important role in the implementation of environmentally friendly strategies. For example, sensors can track and trigger appropriate adaptive responses in building envelopes to changing external conditions. In addition, smart materials capable of storing heat and even regenerating themselves over time can be used for the optimization of natural resources.

The case studies presented embrace different strategies for thermoregulation in both hot and cold environments. The project modeled on the side-blotched lizard combines behavioral and physiological strategies to provide a comfortable living unit in the fluctuating daily temperatures of the desert. The unit accumulates, stores and later releases heat. Polar bears live in the arctic, one of the most inhospitable environments, yet they can insulate themselves efficiently with the help of their thick fur and blubber as well as by burying themselves in compact dens. The related project was designed to take advantage of the insulating properties of the Earth. Its adaptive envelope tracks the sun seasonally, absorbing as much heat as possible to then be released into the compact interior. The snow leopard lives at high altitudes in the Himalayas yet it minimizes heat loss with its thick fur and respiratory system. The project is a relocatable researcher's laboratory, which has an expandable system that opens up to thermally protect researchers in this extreme region.

Animal Examples — Heat Dissipation   Animals in hot environments are often hairless to increase heat dissipation. Birds do not sweat but some species, such as storks, defecate on their featherless legs to increase heat loss via evaporative cooling. Camels allow their body temperatures to rise during the day, drawing heat away from the vital organs and storing it in their fatty humps. The large ears of the African elephant and its hairless body also increase the potential to dissipate heat. Many desert animals, such as the tarantula and fennec fox, are nocturnal, spending the hot days in cool burrows and emerging in the cold night. The deep beak of the toucan is primarily used for feeding, but also contains a network of blood vessels through which body heat can be radiated to the environment. Large reptiles such as alligators and crocodiles may bask in pools of water during the hottest parts of the day to avoid overheating.

Animals use physiological and behavioral mechanisms to dissipate heat in hot environments. (Photographs courtesy of: yellow-billed storks, B. Larison; Arabian camel, S. Yeliseev; African elephant, A. Kirschel; fennec fox, T. Parkinson; toucan, J. Drury; tarantula, M. Hedin; American crocodile, M. Tellez)

Animal Examples — Heat Conservation Animals living in cold environments, such as penguins and walruses, tend to be larger than their relatives in more temperate environments to minimize their relative surface area. Layers of insulating fat can also keep the body warm. Cold-blooded animals have different approaches to keeping warm. Invertebrates such as butterflies must bask in the open during the warmest part of the day to heat up. Snowy owls and penguins also possess fine, downy feathers that can fluff to trap warm air close to the body. Musk oxen and sea otters have dense underfur for insulation and water-repellent guard fur to prevent snow or cold water contacting warm skin. Some fish living in arctic temperatures, such as the Amur bitterling, produce antifreeze-like chemicals that circulate in their bloodstream and prevent ice crystals from forming.

Animals have many ways to gain or preserve heat, particularily in cold environments. Some can live in extremely harsh conditions. (Photographs courtesy of: king penguin, morgueFile.com; butterfly, S. McCann; walrus, Wikimedia; snowy owl & sea otter, S. McCann; musk ox, J. Bussey; Amur bitterling, S. Yeliseev)

Animal Examples Reptiles have several behavioral thermo-regulatory mechanisms. Collared lizards prefer to keep their body temperatures at elevated levels and bask in the sun in open areas to gain heat. On very hot days they raise themselves off the ground to prevent gaining heat from radiation. Gopher tortoises dig several very long burrows underground to protect themselves from heat, losing moisture and predators. They spend most of their lives in these burrows. Like collared lizards, eastern fence lizards are often found in open habitats in the sun. They are active predators and keep body temperatures at an optimum for locomotion and digestion. Sidewinder rattlesnakes get their name from the way they move across the desert sand. They change from being active during the day in the cooler months to being nocturnal in the hottest months.

Many reptiles inhabit deserts. Their activity patterns can be modified to prevent them from overheating. (Photographs courtesy of: collared lizard & sidewinder rattlesnake, P. Niewiarowski; gopher tortoise, S. McCann; eastern fence lizard, D. McShaffrey)

Side-blotched lizard
Uta stansburiana

Phylum:	Chordata	Family:	Phrynosomatidae
Class:	Sauropsida	Genus:	*Uta*
Order:	Squamata	Species:	*U. stansburiana*

Photograph courtesy of D. McShaffrey

Habitat & Climate   Lizards in the genus *Uta* are known as side-blotched lizards and are common inhabitants of arid regions in the western United States. There are several species of *Uta*. The common side-blotched lizard, *Uta stansburiana*, occurs south to north from central Mexico to Washington and east to west from western Texas to the Pacific coast. The species prefers open habitats in deserts and scrublands with elevations ranging from 0 to 2500 m (0–8200 ft). The project's location, the portion of the Great Basin Desert within Utah, is characterized by its northern latitude and high elevations of 900–2000 m (2953–6560 ft), making it cooler than other deserts with temperatures ranging from −18°C (−0.4°F) in January to 50°C (122°F) in July. It rains mostly during the months of November and April, with an average annual precipitation of 2.5 cm (1 in.). High winds occur during the whole year but become stronger during the late winter and early spring.

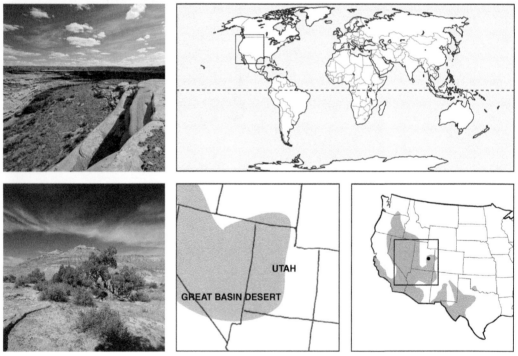

The topography of the Great Basin Desert includes flat valleys interspersed with small mountain ranges. Rocks provide places for basking in the sun as well as places to hide from numerous predators. Sagebrush plants are common and provide shade in the hot hours of the day. Additionally, the desert basin with eroded terraced cliff sides provides shaded overhangs. (Photographs courtesy of D. McShaffrey)

Animal Physiological, Behavioral & Anatomical Elements
Lizards are ectotherms, meaning they use their environment to regulate their body temperature. Temperature regulation in lizards is achieved through a combination of their skin characteristics and their behavior. The common side-blotched lizard has a skin coloration pattern that is typically a dark color on the back for sunlight absorption and a light color on the abdomen to reflect heat from the ground.

Lizards obtain or dissipate heat from the environment through their behaviors. A lizard will adjust its body position to be perpendicular to the sunlight for heat absorption or parallel to the sunlight while curling up its toes to avoid heat gain by minimizing the area of the body touching the ground. They spend the hottest hours of the day in the shade to prevent overheating.

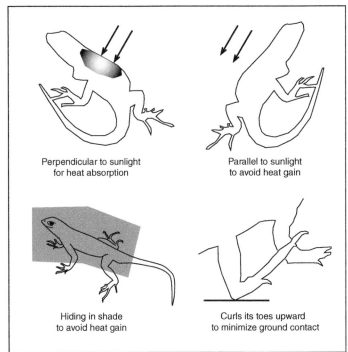

Perpendicular to sunlight
for heat absorption

Parallel to sunlight
to avoid heat gain

Hiding in shade
to avoid heat gain

Curls its toes upward
to minimize ground contact

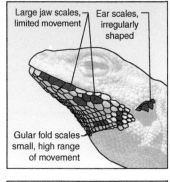

Large jaw scales,
limited movement

Ear scales,
irregularly
shaped

Gular fold scales
small, high range
of movement

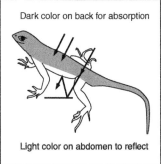

Dark color on back for absorption

Light color on abdomen to reflect

The side-blotched lizard has many behaviors that help its ability to thermoregulate in the desert climate: standing perpendicular to the sun to absorb heat, standing parallel to the sun to avoid its rays, curling its toes upwards to minimize contact with the hot floor, and moving to shade when it is too hot. (Photograph courtesy of D. McShaffrey; project team: J.M. San Pedro, A. Nahmgoong, & Y. Yuan)

Interface between the Skin & External World Lizards' skin scales are important for camouflage, preventing desiccation, and protection against sharp rocks and foliage. The skin consists of two principal layers: the epidermis, the outer layer, and the dermis, the inner layer. Lizards' scales are thickenings of the epidermis and are primarily made of a horny substance called keratin, much like human fingernails.

The scales are thickenings of keratin, connected by hinges of thin keratin; they are often folded and overlap each other. Areas of the body with smaller movement have smaller scales to allow for increased flexibility; large scales are found on areas of the body with restricted movement. The outer skin is molted periodically and then renewed by cells in the inner skin layer. Molting allows room for growth and at the same time replaces worn-out skin.

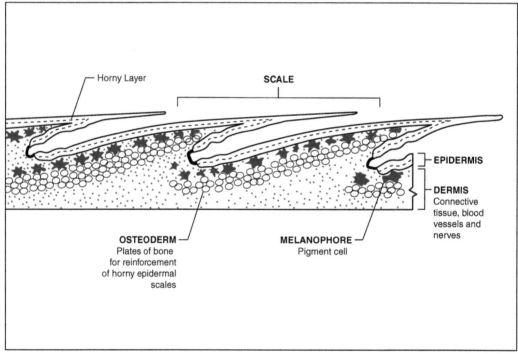

Lizard scales are one continuous keratin surface formed from the epidermis; they consist of thickened regions (the horny layer) and thin, connecting hinges. The dermis contains osteoderms and pigment cells (melanophores). The different sized scales contribute greatly to the lizard's mobility and its ability to thermoregulate. (Photograph courtesy of D. McShaffrey)

Interrelationship between the Skin & Internal Systems   The common side-blotched lizard's physiology is optimized for efficiency in extreme temperatures and in water-scarce conditions. They are opportunistic feeders, readily feeding on invertebrates, such as mealworms, arachnids or insects, living in their habitat. Their digestive system absorbs all available water from their prey. The excrement they produce is a dry white paste, which also serves as a water conservation strategy.

In addition to aiding with thermoregulation and camouflage, throat coloration in *U. stansburiana* reflects a type of mating strategy observed in males. Males have orange, blue or yellow throats which correspond to their level of aggression and ability to maintain pair bonds with females.

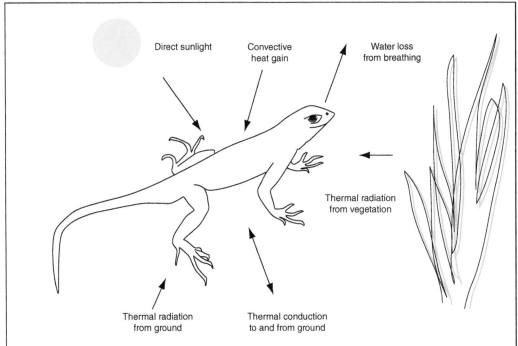

The common side-blotched lizard faces many different sources of heat in the desert. Despite these sources, it has adapted a physiology and behavioral repertoire that allows it to thrive in hot and dry environments. (Photograph courtesy of D. McShaffrey)

Proto-Architectural Project Thermoregulation requires constant vigilance for desert-dwelling organisms, particularly ectotherms. The project investigates how a building envelope system could respond to the extreme temperature variations of the desert in order to keep the interior temperature within human comfort levels. The project is a small artist's retreat located in the Great Basin Desert inspired by *U. stansburiana's* physiological and behavioral adaptations to the desert temperature regime.

The design's focus is to create an enclosure system which maintains 24 hour comfort for the residence during hot, arid days and sometimes very cold nights. The physiological strategies of the lizard's skin have inspired the house walls' design, while the lizard's behavioral adaptations have influenced the "smart" sun-tracking system actuated by a hydraulic system and sensors.

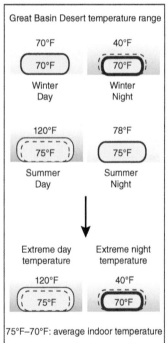

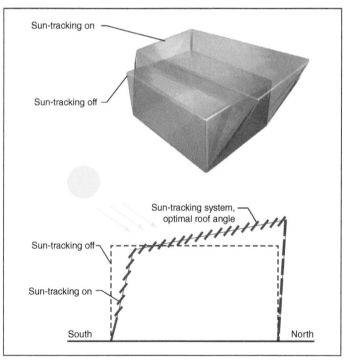

S.C.A.L.E.S. (smart – continuous – active – layered – environmental – system) is the culmination of the observed efficient thermoregulation of the lizard; it combines the characteristics and behaviors that help it survive in the desert, and integrate it in the building envelope. It takes cues from survival skills of the lizard and makes the building, in essence, survive in the desert quite comfortably.

Project Documentation S.C.A.L.E.S. uses modular panels, distributed throughout the entire envelope, in different ways, depending on their functions. The south-facing wall is composed of three different types of panels: opaque insulative, photovoltaic, and operable window. The insulative panel uses phase change material to allow for a stable interior temperature throughout the day. The panel is hollow and filled with a bio-based phase change material. Heat gets stored during the day, while keeping the interior cool. The heat collected during the day is slowly released and heats the residence at night.

The envelope's structural system is made up of a braced steel grid to which the panels attach. All façades follow a similar organizational strategy, while the panel composition may vary depending on their exposure.

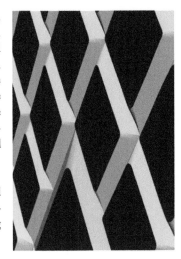

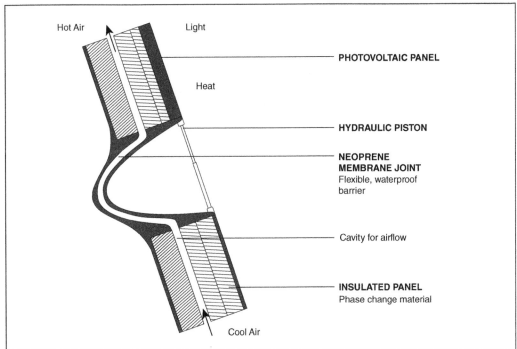

The photovoltaic panel captures the sun's rays and converts them to the studio's electricity. The window panel allows for views and ventilation. These panels are strategically arranged to maximize their performance. Between the panels is a flexible, foamed neoprene membrane that allows the panels a range of motion, controlled by sensors, while being continuous, insulative, and waterproof.

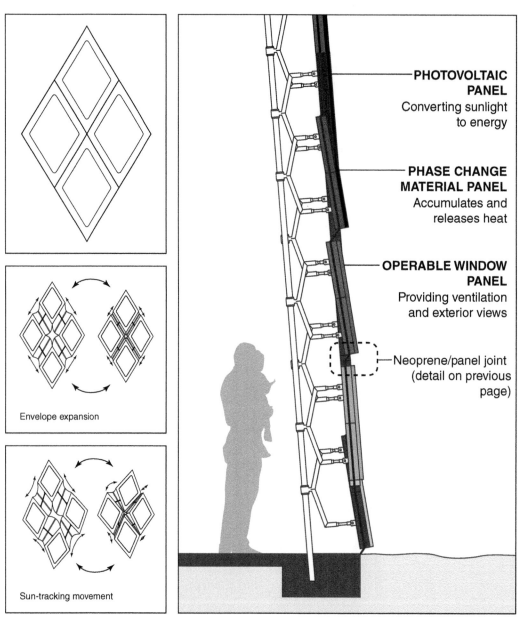

PHOTOVOLTAIC PANEL
Converting sunlight to energy

PHASE CHANGE MATERIAL PANEL
Accumulates and releases heat

OPERABLE WINDOW PANEL
Providing ventilation and exterior views

Neoprene/panel joint (detail on previous page)

Envelope expansion

Sun-tracking movement

The wall is composed of rhomboidal panels mounted on hardware that allows for a small range of movement. The scales are mounted on universal joints, gas springs, and rod ends. The universal joints allow for the panel to tilt horizontally and the gas springs move opposite to one another and allow for vertical tilt.

The artist studio fixes itself directly on the desert floor, much like the lizard. The individual panels gleam in the sunlight while collecting heat and energy. The structure maintains a slight tilt in the roof and south façade in order to be optimally positioned to the sun angle.

Animal Examples Large ungulates have several adaptations to allow them to inhabit cold climates. Moose and bison grow dense fur in the winter to maintain body heat. They also reduce their metabolism and lower their activity levels to conserve energy in the winter. Alpacas, along with their thick coats of fur, have behavioral mechanisms to conserve heat. They protect areas of their bodies with short fur or thin skin in extreme conditions.

Ermines, or short-tailed weasels, replace their brown summer coats with a thick white coat of fur in the winter. This coloration helps to camouflage them in the snow to avoid predation and to catch prey. They also use dens in the snow to protect them from the elements.

Many mammals have dense winter coats that allow them to preserve body heat in the winter. (Photographs courtesy of: moose, B. Miers; American bison, D. Greenfield; alpaca, J. Drury; ermine, L. Parenteau)

Snow leopard
Panthera uncia

Phylum:	Chordata	Family:	Felidae
Class:	Mammalia	Genus:	*Panthera*
Order:	Carnivora	Species:	*P. uncia*

Photograph courtesy of D. Conner, Snow Leopard Trust

Habitat & Climate   There are five species in the genus *Panthera*, commonly referred to as the "big cats." Snow leopards inhabit the mountainous regions of central and southern Asia, their range covering an area of nearly 2 million km² (772,205 mi²). Snow leopards usually reside at altitudes between 3000 and 5400 m (~9842–17,717 ft). Here, the environment is harsh and forbidding, with temperatures ranging from −30 to 15°C (−22 to 59°F).

The climate is cold and dry with only small amounts of precipitation during the summer months. As a result, the mountain slopes are sparsely vegetated. Snow leopards prefer these rocky slopes as they provide a good overview and cover to help them sneak up on prey. The harshness and remoteness of these habitats means that it is difficult to study snow leopards in their natural environments.

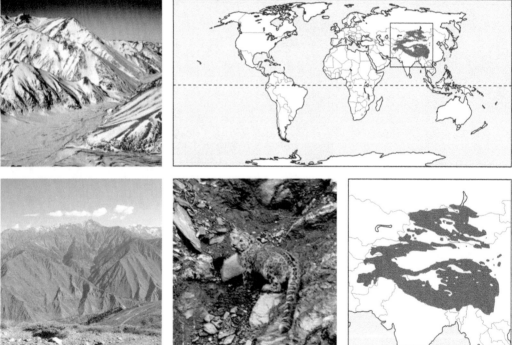

The snow leopard's range in the rugged mountainous regions of central Asia stretches through twelve countries: Afghanistan, Bhutan, China, India, Kazakhstan, the Kyrgyz Republic, Mongolia, Nepal, Pakistan, Russia, Tajikistan, and Uzbekistan. The home range of a single animal varies from 35 to 1000 km², depending on prey availability. (Photographs courtesy of: snow-covered peaks, Snow Leopard Trust; top image by B. Hogue; landscape and leopard, Felidae Conservation Fund)

Animal Physiological, Behavioral & Anatomical Elements
Adults measure around 60 cm (~2 ft) at the shoulders, 1.8–2.3 m
(~6–7.5 ft) between the nose and the tip of the tail and weigh
35–55 kg (~77–121 lbs.). Males are generally 1/3 larger than
females. Most features of the snow leopard relate to thermoregulation. Compared to other cats, their bodies are sturdier and
rounder to minimize surface area and prevent heat loss. Their
small, rounded ears are covered in thick fur. Short forelegs, long
hind legs, and powerful chest muscles allow snow leopards to rapidly pursue prey over unstable terrain. An enlarged nasal cavity
and lung capacity compensates for low oxygen levels at high altitudes. Cold, dry inhaled air is warmed and moistened as it passes
over the delicate tissue covering the nasal turbinate bones, which
also collect heat and moisture with every exhaled breath to help
preserve body temperature and retain valuable water.

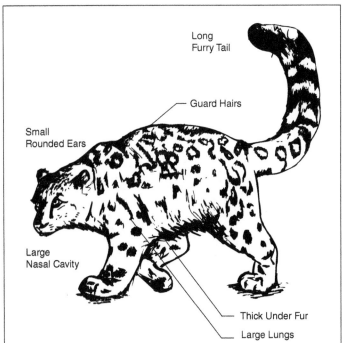

The snow leopard uses its long, thick tail for balance when pursuing prey over the mountainous terrain. Densely furred paws allow movement across snow while the stocky body and thick fur coat prevent heat loss and insulate in subzero temperatures. (Photographs courtesy of: snow leopard body and face, Snow Leopard Trust, face by F. Polking, body by M. Trykar; paw, Felidae Conservation Fund; project team: J. Hoen & G. Mamoune)

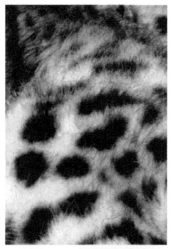

Interface between the Skin & External World Snow leopards are covered with dense, insulating fur. Hairs on the belly can be up to 12 cm long. The coat is a white color with brown or gray-brown spots during winter, darkening to a light gray color with dark brown to black spots during summer. These colors provide superb camouflage all year round. Spots are mainly located on the animal's back and are formed by straight guard hairs, which are up to 70 mm long and between 40 and 60 μm in diameter. The latter is composed of soft and slightly curved, nonmedullated hairs with a thickness between 5 and 20 μm. This fine layer of hairs provides most of the thermoregulating properties of the snow leopard's fur. The thick tail, which is as long as the body, is used primarily as a counter-balance during climbing and chasing. However, it can also be used as a muffler to cover the face when resting.

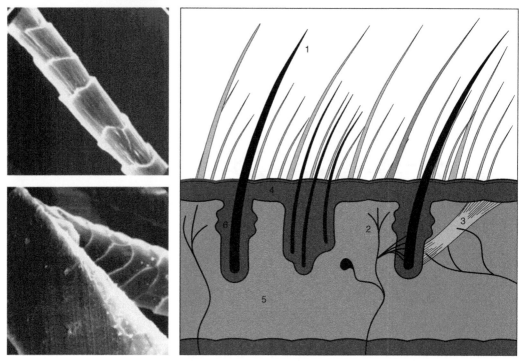

Microscopic images of the cuticlar structure of the fine fibers and a transverse section of an intermediate fiber. The snow leopard does not sweat through the skin, which means the thermoregulating properties of the fur are solely of an insulating nature. Cross-section of skin: 1. Guard hair, 2. Temperature and pain receptors, 3. Hair erector muscle, 4. Epidermis, 5. Dermis, 6. Hair follicle (Photographs courtesy of: leopard cub, Snow Leopard Trust, by H. Freeman; hair micrographs, A. Galatik)

Interrelationship between the Skin & Internal Systems While fur provides a layer of insulation for the skin, cold air still enters the body via the lungs due to the necessary act of breathing. The lungs of the snow leopard are enlarged relative to those of a similarly sized cat in order to maximize oxygen uptake from the thin, high altitude air. Cold air would chill the body and dry the lungs, however, reducing overall respiratory abilities.

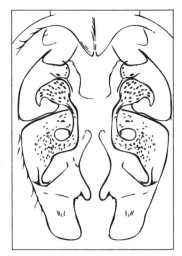

The nasal turbinate bones housed inside the snow leopard's nasal cavity are covered in moist, highly vascularized tissue called respiratory epithelium. This surface warms and moistens inhaled air passing to the lungs and reduces cooling of core temperature. The turbinate surface cools as cold air is inhaled but can trap excess heat and moisture from exhaled air, recouping some energy and reducing wasted water.

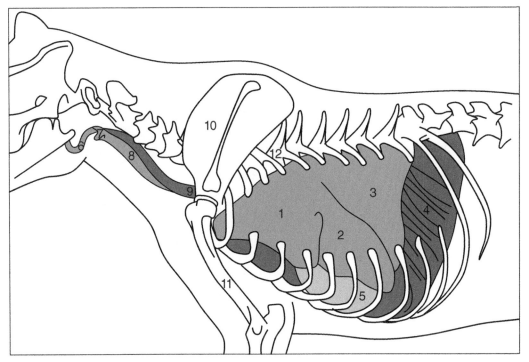

In order to live in extreme altitudes, snow leopards have developed large, powerful lungs capable of extracting enough oxygen for an active life in a thinly aired environment. The nasal chamber (top image) acts as a two-way air conditioning unit. Anatomical diagram (bottom): 1. Cranial lobe of the lung 2. Middle lobe of the lung 3. Caudal lobe of the lung 4. Diaphragm 5. Heart 6. Hyoid apparatus 7. Larynx 8. Trachea   9. Esophagus 10. Scapula 11. Humerus. Bottom diagram redrawn from Atlas of Veterinary Clinical Anatomy, 2004)

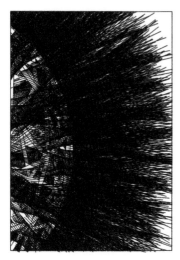

Proto-Architectural Project Little research has been achieved on the snow leopard in its natural habitat due to political problems, steep elevation and extreme weather. This project is intended to make their habitats more accessible to researchers through the design of a structural envelope that could be flown in and deployed on mountainsides. This structure serves as a base and shelter for researchers wanting to perform field studies of the cats. The main responsibility of the building envelope is thermoregulation, mainly protecting the interior from exterior temperatures varying from −30 to 15°C (~ −22 to 59°F). Protection from cold, harsh weather is important for survival in these mountainous regions, and will only grow in importance worldwide as global warming causes considerable climatic changes. The snow leopard's thick fur, body shape, and respiratory system are prime examples demonstrating nature's ability to deal with a cold environment.

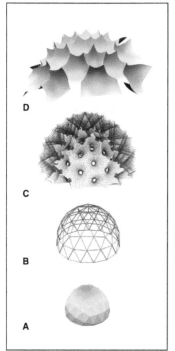

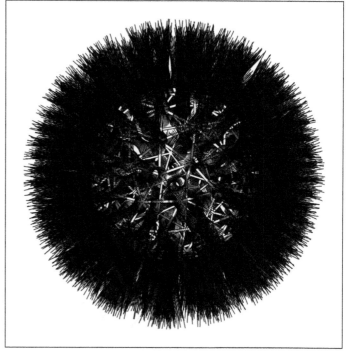

A: Main structure insulated by vacuumed insulated panels. B: Structure forming the wire frame of a geodesic dome and made expandable by telescopic members. C: Clusters of rods fixed to each other and a base causing them to open and close as the skeleton underneath expands or contracts. D: Dark colored, waterproof membrane. By emulating muscular features of the snow leopard, the proposed structure expands or contracts to create insulating pockets within the envelope.

Project Documentation   The multilayered envelope emulates the anatomical interactions between the animal's systems. The inner static structure is based on a geodesic semi-sphere with highly efficient vacuumed insulated panels, which function as the innermost layer of insulation — expanding and contracting as the frame moves — and a main structure for the rest of the system to build upon. A system of telescopic rods, mounted on the inner body, forms a geodesic dome that pneumatically expands and contracts. The structure's connection points contain a round platform hosting a number of hinged and pivoting rods, connecting one end to the platform and the other end to a corresponding rod. When the telescopic structure expands, the platforms move apart, forcing the rods to flatten and open each cluster. A final dark-colored waterproof membrane stretched over the end of the rods unfolds, heating the enlarged air pocket underneath through solar heating.

Taking cues from the erector muscles in the snow leopard's skin, the expansion and contraction of the geodesic structure cause each cluster of rods to open or close according to external climatic conditions. This happens by connecting each rod to a corresponding rod in an adjacent cluster. When the distance between the cluster origin points increases each cluster is opened.

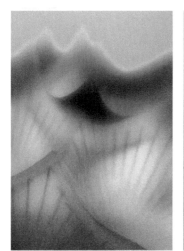

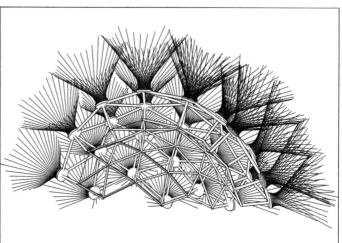

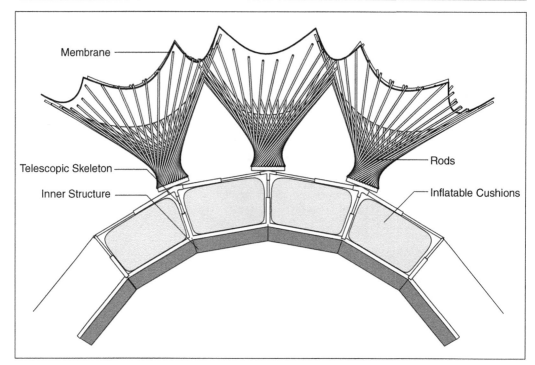

Membrane

Telescopic Skeleton

Inner Structure

Rods

Inflatable Cushions

Close up of façade. An axonometric view of the envelope depicting the geodesic structure and the platforms forming each cluster of rods. In a typical section through the building envelope the movement in the envelope is ensured by a series of interconnected inflatable cushions which eliminate the need for local mechanical actuators in each node.

The project's formal expression, shaped by the tensioned membranes, inserts itself within the skyline of the mountain peaks present in the surroundings in which the research facility is located. The pod is a rather curious object in this environment with its repetitive unified elements and materiality.

Animal Examples   The cold provides formidable challenges for mammals. Seals and other marine mammals must endure low temperatures almost constantly. Large body masses, combined with a layer of insulating fat called blubber, help maintain warm core temperatures.

For terrestrial mammals, such as hares, a layer of thick fur helps insulate the body. Mammals living in extremely cold environments may evolve additional adaptations to the cold. Arctic foxes have short ears and limbs to reduce their body surface area and minimize heat loss. A white coat additionally provides camouflage in the snow. The Canadian lynx hunts on deep snow. Broad, fur-covered feet act as snow shoes, allowing the lynx to run over the loose snow surface.

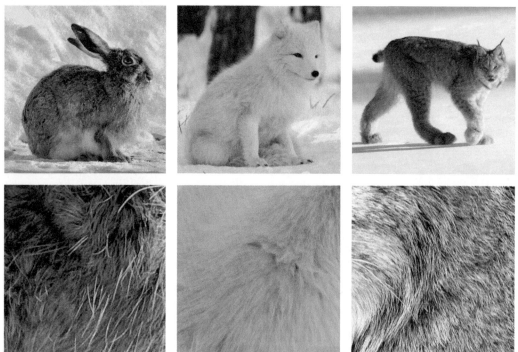

Some adaptations that allow animals to thrive in cold conditions are thick fur, a layer of blubber, and reduction of body surface area. (Photographs courtesy of: elephant seal, J. Himes; arctic hare, A. Holmberg; arctic fox, T.S. Bortne; Canada lynx, K. Williams)

Polar bear
Ursus maritimus

Phylum:	Chordata	Family:	Ursidae
Class:	Mammalia	Genus:	*Ursus*
Order:	Carnivora	Species:	*U. maritimus*

Photograph courtesy of M. Johnson

Habitat & Climate   The family Ursidae consists of eight living species. The polar bear is a particularly distinct species, which is found throughout the Arctic Circle, across North America, Europe and Asia. Polar bears are often characterized as marine mammals as they are comfortable in and out of water and are dependent on ice as a platform for hunting. The focal region for this project is the Hudson Bay in Canada.

In this region the average air temperature in summer is 10°C (50°F) and the minimum winter temperature reaches −23.5°C (−10°F). Winters are long and dark, with a maximum sun angle of 6° while summers have long days with a sun angle that reaches 56°. Some arctic areas are covered with ice all year long, while others may lose ice, allowing the arctic tundra vegetation to bloom.

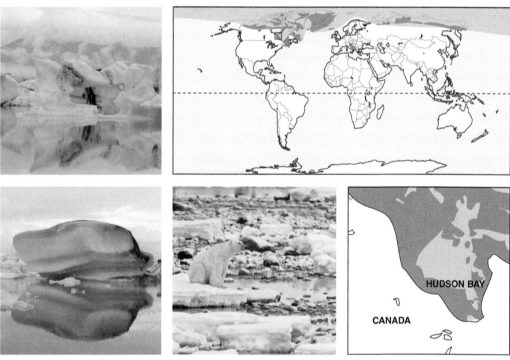

Polar bear distribution includes the U.S. (Alaska), Canada, Russia, Denmark (Greenland), and Norway. Polar bears spend most of their lives on ice. Their range spans the arctic regions of all continents where they roam pack ice searching for seals. They venture on land only during the summer, when the ice melts. (Photographs courtesy of: sunset and polar bear, M. Johnson; glaciers, G. Rochette, morgueFile.com)

Animal Physiological, Behavioral & Anatomical Element
Polar bears are extremely large. Males can measure up to
3 m (~10 ft) in length and weigh an average of 500 kg (1100 lbs).
Females may weigh half as much as males. The large body mass,
combined with short limbs and ears, reduces the surface area to
volume ratio of the body preventing loss of body heat.

Thick fur and a layer of insulating fat provide the primary means
of insulation against the harsh arctic cold. Polar bears hibernate
through the winter in dens dug into snow banks or the ground.
Polar bears roam vast distances over the sea ice in summer
searching for seals. Large, broad paws and sharp claws provide
snowshoe-like assistance when walking over snow. They are
excellent swimmers and have been observed to stay in the water
for prolonged periods of time when moving between ice floes.

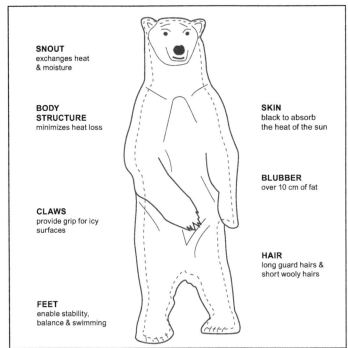

SNOUT
exchanges heat
& moisture

**BODY
STRUCTURE**
minimizes heat loss

SKIN
black to absorb
the heat of the sun

BLUBBER
over 10 cm of fat

CLAWS
provide grip for icy
surfaces

HAIR
long guard hairs &
short wooly hairs

FEET
enable stability,
balance & swimming

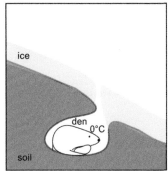

ice

den 0°C

soil

The polar bear is the largest living carnivore. Its entire appearance reflects life in the cold, hostile arctic. Newborns
are less well protected and are born in dens dug into snow or earth. They emerge in summer, already able to tolerate subfreezing temperatures. (Photographs courtesy of: snout and mother with cub, M. Johnson; project team: im
studio mi/la & I. Mazzoleni)

Interface between the Skin & External World   Arctic temperatures can reach −45°C (−49°F) during the depths of winter. The polar bear's primary defense against the cold is its skin. Externally, polar bears are insulated by a thick, white fur coat that covers almost the entire body, including the soles of the feet. The fur is composed of two layers. The dense white underfur provides the main source of insulation by trapping warm air close to the skin and, conversely, by preventing contact by ice and water. In the outer layer, longer guard hairs are hollow and transparent, lacking pigments which also act as camouflage in the snow. Polar bears molt during the summer. Beneath the skin within the hypodermis, polar bears are further insulated by a layer of fat. Additional features, such as black skin, particularly on exposed areas such as the nose and tongue, allow the bears to absorb solar energy and prevent additional passive heat loss.

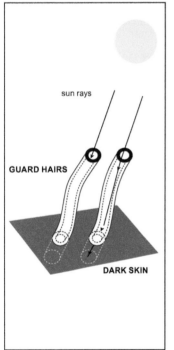

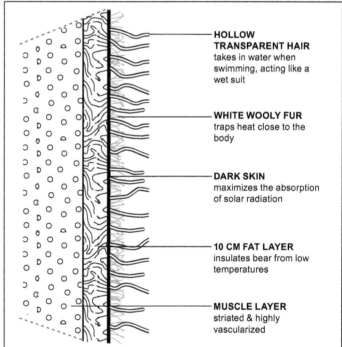

HOLLOW TRANSPARENT HAIR takes in water when swimming, acting like a wet suit

WHITE WOOLY FUR traps heat close to the body

DARK SKIN maximizes the absorption of solar radiation

10 CM FAT LAYER insulates bear from low temperatures

MUSCLE LAYER striated & highly vascularized

Fur is the polar bear's primary defense from the harsh cold. The dense underfur traps warm air close to the skin, warming the body. Longer, hollow guard hairs protect the underfur from the elements. The skin is also effective in providing protective layers that are capable of trapping water as the bear swims in order to reduce heat loss. Beneath the skin, a thick layer of fat provides additional insulation. It has been proposed that the hollow hairs guide UV radiation to the skin, similar to fiber optic cables, but this now seems unlikely. (Photograph courtesy of D. McShaffrey)

Interrelationship between the Skin & Internal Systems   Like all mammals, polar bears are endothermic, meaning that they produce their own body heat internally. Their large body mass, combined with short extremities, thick insulating blubber, and fur, make the polar bear exceptionally efficient at retaining body heat. In fact, overheating becomes problematic for polar bears, even in subzero temperatures. This prevents polar bears from undergoing prolonged running. Swimming or rubbing themselves in snow are efficient ways of cooling down.

Viewed through an infra-red camera, polar bears are almost invisible; only the warm air leaving the nose and superficial blood flow to the less furred head provide signs of life. Also, the blood cells are adapted to hold more air, allowing for longer stints under water.

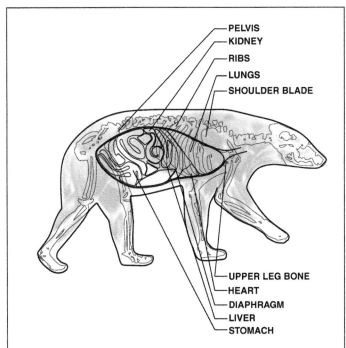

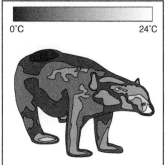

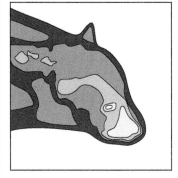

The polar bear's large body mass reduces its surface area to volume ratio, which helps to retain body heat. With thick fur, polar bears are so efficient at retaining heat that they appear almost invisible to infra-red cameras; only warm air lost via the nose can be distinguished. (Photograph courtesy of D. McShaffrey; anatomy diagram adapted from National Bowhunters Education Foundation [2006]; thermal diagrams adapted from P. B. Dill and L. Irving [1964])

Proto-Architectural Project This project mimics the physiological and behavioral adaptations which have allowed the polar bear to survive in the planet's harshest weather conditions. The project attempts to optimize natural local resources to provide energy needed to regulate human thermal comfort. The compact living units are partially embedded in the earth, similar to the bear's hibernation den. The units' southwest orientation optimizes the heat gain from the sun. The sun's energy, heat and light, is harvested by the active building envelope composed of hollow re-orientable fur-like glass tubes and travel to the insulating strata where the energy is stored, conserved and slowly released into the compact pod. The phosphorescent cells embedded in the phase changing material (PCM) collect light, which is slowly released at night creating an atmospheric sky-like vaulted ceiling.

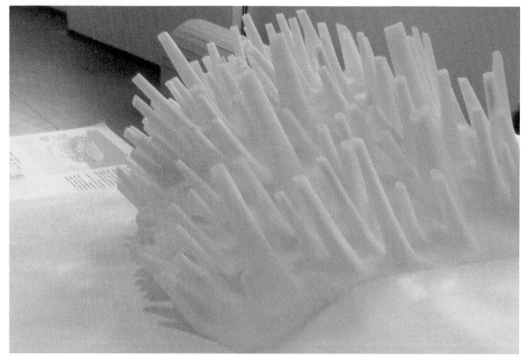

Studies of optimized shapes were facilitated by computer modeling to aid in understanding and generating complex geometries. Computational/digital studies were further developed through testing with rapid prototyping and by engaging three-dimensional drawings to further explore the inherent spatiality. These preliminary investigations are necessary to proceed in the development and fabrication of full-scale mockups to test the feasibility of current ideas and material strategies.

Project Documentation   The envelope is a storage for heat and light which minimizes the use of active systems and maximizes the use of passive systems. The storage of heat and light within the active envelope varies according to seasonal changes. During the winter, light and heat are transferred into the unit, where heat is conserved within building-integrated thermal storage in the form of PCM. During the long, cold summer nights, light is managed and minimized while the heat is still being collected. This variability in conveyance and conservation of heat and light is accomplished via sensors within the movable tubes which track the sun angle and optimize the heat accumulation. The thick envelope assembly controls a number of crucial factors, including thermo-physical properties of the materials, the outdoor climate, and the operating schedule of the compact dwelling.

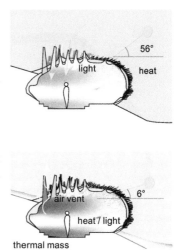

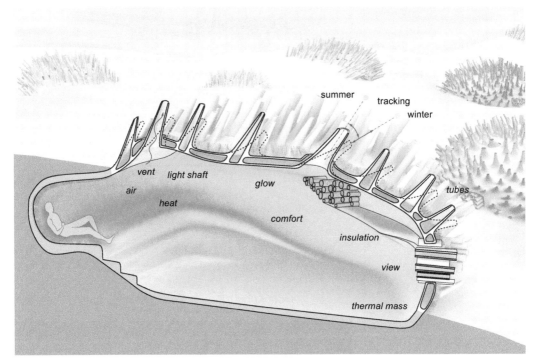

The dwelling is articulated around a compact central space to avoid energy dispersion. Each area satisfies multiple programmatic functions. For example, the lounging area is the living space as well as the sleeping space, while the steps allow for small gatherings to observe the sky by looking through the scattered openings on the ceiling.

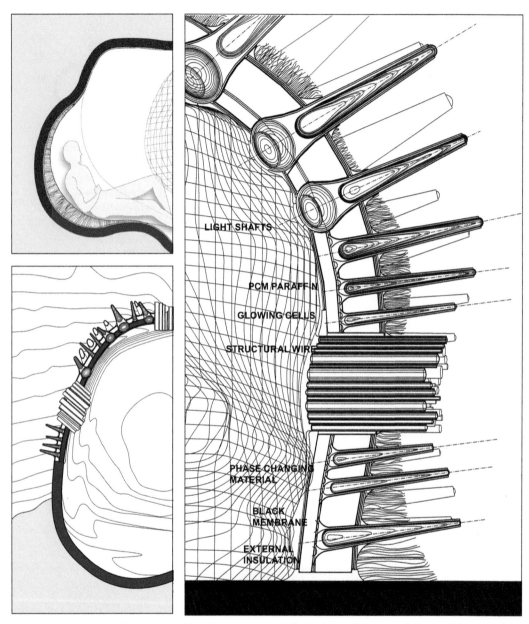

Labels in figure: LIGHT SHAFTS, PCM PARAFFIN, GLOWING CELLS, STRUCTURAL WIRE, PHASE CHANGING MATERIAL, BLACK MEMBRANE, EXTERNAL INSULATION

This structure is a super-efficient habitable cell. Its openings, diffusely directed, capture all the possible energy from the sun, accumulating light and heat and dispersing them through high-tech materials and technologies. In exploiting the available energy of the sun, this unit is a comfortable shelter, welcoming life and providing protection from the relentless environment of the extreme north.

The cold that penetrates the skin deeply inside. The reverberation of the light on the white flat polar landscape, so strong, almost unbearable. The horizon so far, feels endless. The lost, bewildered look expresses the void. All of a sudden, something is delineated, something starts to reveal itself: the vision, the imagination of something, perhaps a glare, some shapes, undefined creatures, feral structures.

5 Water Balance

Amphibians such as frogs have a close relationship with water. Their skin is permeable, and many stay moist by living in or near water. (Photograph courtesy of S. McCann)

Life on Earth has an intimate relationship with water. All organisms, from humans to bacteria, are primarily composed of water; it is the main component of our cells, and most of our tissues and organs are bathed in it. It is through water that nutrients, vitamins and hormones are delivered to different parts of the body where they are used for producing energy and carrying out life's basic functions. Different environments place strong demands on water use and, as such, animals exhibit many diverse adaptations for collection and conservation of water and protection from being exposed to too much water. On one extreme, organisms in harsh environments like the desert struggle to acquire enough water to survive, while on the other extreme aquatic organisms are surrounded by water and must prevent their tissues from becoming overcome by it. While different organisms have different relationships with water, maintaining a balance of water within the body is crucial for survival.

Water cannot be wasted in the desert environment, because it can only be acquired from a few sources. Most desert creatures obtain water from plants, particularly those that are adapted to store water, like cacti. Others meet all of their water needs from the food they eat. Desert-adapted animals have evolved innovative strategies to gain water and prevent water loss. Some retain water in fatty tissue within the body; when the fat is metabolized the water in the tissue is released and can be used by the animal. Reptiles and birds excrete waste in the form of uric acid, which wastes less water than the urea excreted by amphibians

Animals with permeable skin use many mechanisms to prevent the loss of water from their tissues. For example, snails and salamanders both secrete mucus in order to stay moist. (Photographs courtesy of: snail, P. Thompson; salamander, H. Thomssen)

and mammals. Many reptiles excrete very dry feces to further preserve water. Desert-dwelling mammals such as kangaroo rats possess highly efficient kidneys and excrete extremely concentrated urine to retain ingested water. Exhaled water vapor may also be recaptured in the nasal cavity to prevent further water loss. Behavioral adaptations such as burrowing underground not only provide a cooler temperature in the day, but can also aid in water balance, because soil is cool and moist.

Though not as extreme as desert conditions, other biomes possess their own set of water conservation challenges that must be overcome by the organisms living there. In all terrestrial habitats the concentration of water in the environment is lower than it is inside the body of animals. Many animals have thick outer skin layers that prevent them from losing water to the environment. Arthropods generally have a waterproof wax within the top layer of their exoskeleton; they can also close pores used for breathing to prevent additional water loss in dry conditions. Reptiles and mammals use the protein keratin in their skin as a barrier to water loss. Mammals may also produce an oily film to cover and waterproof their skin and fur. The feathers of birds lock together and align in such a way as to confer waterproofing.[15] All cells use salt ions, which are charged particles, to function. As the ions flow in and out of cells, they create electrical gradients across the cell membranes. These gradients are integral to all aspects of cell function, from transporting sugars for energy, to firing of neurons that transmit information from the brain to muscles. Maintenance of the concentration of salts to water, or osmoregulation, is critical for most animals. Animals that live in or have a close association with water do not have an issue with preventing water loss; however, they do need to keep the concentration of molecules in water and the concentration of molecules inside their bodies balanced.[16]

Architecture has always been concerned with issues related to humidity control and moisture penetration, and a lot can be learned from nature when it comes to water management. Water balance in wet climates has traditionally focused on keeping out rain and humidity, while in arid climates architects have learned to add humidity to dry air. In traditional Islamic architecture, for example, the insertion of small fountains in courtyards provides comfort by means of evaporative cooling. In tropical climates strategies to achieve comfort are achieved by optimizing natural ventilation and detachment of the structure from moist soil.

The Alhambra region is rich in fountains and courtyards for evaporative cooling to cool the rooms surrounding the patio. Patio de los Arrayanes, Alhambra, Granada, mid 14th century. (Photograph courtesy of L. Pretorius)

Court of la Acequia. Palacio de Generalife, Alhambra, Granada, 1309. (Photograph courtesy of L. Pretorius)

Today, water conservation has become a fundamental driver of design innovation throughout the world in order to lessen our environmental impacts. Increasing population sizes have led to the establishment of new settlements in areas of the planet where humans struggle to find the basic resources required for sustainability, such as water. Strategies for water conservation can be developed at many design scales, ranging from low-flow plumbing fixtures to irrigation systems, and include both collection in cisterns and the reuse and treatment of gray and black water.

An understanding of water balance in the banana slug, with its permeable, mucus coated skin, led to the design of a greenhouse that collects water in manmade bladders (or cushions) that provide irrigation to plants. The tropical environment of the dyeing dart frog inspired the development of museum walls designed to remove moisture from the air to protect valuable artifacts. The ochre sea star's ability to live both under and above water motivated the design of a retractable structure growing under a pier. Namib desert beetles' adaptations to extremely harsh environments inspired the design of a research facility that, by capturing fog, provides water for human use.

Animal Examples — Water Balance Marine invertebrates such as sea slugs, jelly fish and anemones possess thin body walls that allow for diffusion of gases and nutrients via osmosis. These animals maintain a salt concentration in their tissues that is the same as that of the surrounding seawater in order to balance the concentration of salt to water in their cells. On land, different strategies are required to prevent water loss. Crabs spend part of their time out of water and possess tough exoskeletons that prevent desiccation. Completely terrestrial invertebrates such as earthworms that lack an exoskeleton are restricted to moist environments to prevent loss of water. Most insects possess an exoskeleton that frees them from this constraint. However, the exoskeletons of the smallest insects, such as springtails, are so thin that they cannot prevent water loss. These tiny animals are therefore restricted to moist environments where the risk of passive water loss is minimized.

Animals that live outside of water have several mechanisms to prevent desiccation. Animals that live in water still have to regulate the balance of salts and water in their cells. (Photographs courtesy of: sea cucumber, A. Jaffe; tadpole, K. Pease; octopus, B. Larison; earthworm, morgueFile.com; jellyfish, J. Himes; crab, J. Drury; springtail, S. McCann)

Animal Examples — Water Balance   Marine fishes such as eels maintain a lower salt concentration in their tissues than in the surrounding water by actively regulating salt levels through the gills and by excreting concentrated urine. Amphibians retain many similarities to their fish ancestors, as indicated by the fully aquatic tadpole stage. Adult amphibians retain thin skins, allowing some gas diffusion that makes them susceptible to dessication. Thus, amphibians are restricted to semi-aquatic habitats, as found in newts or caecilians burrowing in moist soils. Whales and dolphins are mammals that have returned to the water – these animals face similar challenges as marine fishes and have a thick, nonporous skin and efficient kidneys for maintaining low blood salt levels. Fetuses are bathed in fluid which provides cushioning to the developing animal, transports nutrients from the mother to the baby, and is the medium through which breathing occurs.

Animals that live outside of water have several mechanisms to prevent desiccation. Animals that live in water still have to regulate the balance of salts and water in their cells. (Photographs courtesy of: salamander, A. Illum; eel, A. Jaffe; pangolin fetus, K. Benirschke; frog & dolphin, H. Thomssen; *Caecilians*, J. Measey; anemone, J. Himes)

Animal Examples Many invertebrates have soft, permeable skin, making them vulnerable to desiccation. Consequently, they often live in moist environments, either in or near water. Turbellaria are a group of mostly free-living flatworms with ribbon-like bodies. Turbellaria have to balance the concentration of salts and water within their bodies, achieved through a specialized system of cells called protonephridia. Land snails keep their skin moist by producing large quantities of mucus and retreating into their shell, which has an opening that can be plugged with a thick layer of mucus while inside. Leeches are segmented worms that primarily live in freshwater habitats such as streams to maintain continuous moisture. Isopods are crustaceans, but unlike other crustaceans they do not live in water. Many isopods have evolved to live in humid habitats, as they breathe through gills that must remain moist.

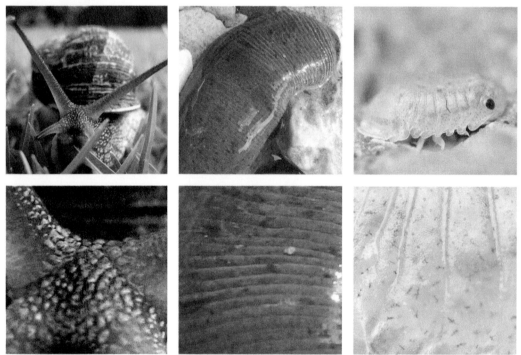

Invertebrates have physiological adaptations to prevent water loss. Turbellaria and land snails have mucus-secreting glands for protection and to stay moist. Isopods excrete waste in a dry form, which helps to conserve water. (Photographs courtesy of: turbellaria, Wikimedia, H. Krisp; land snail, J. Robinson; leech, Wikimedia, C. Schuster; isopod, Wikimedia, G. San Martin)

Banana slug
Ariolimax columbianus

Phylum:	Mollusca	Family:	Ariolimacidae
Class:	Gastropoda	**Genus:**	*Ariolimax*
Order:	Geophila	**Species:**	*A. columbianus*

Photograph courtesy of D. McShaffrey

Habitat & Climate There are three species of banana slug: *Ariolimax californicus, A. columbianus* and *A. dolichophallu.* The California banana slug, *A. californicus,* is the focus of this project and lives on the forest floors of the Pacific coastal rainforest belt, in foggy and forested habitats. The climate of this area is regulated by its proximity to the Pacific Ocean, resulting in a Mediterranean like climate. As a result, the average high temperature ranges from around 15°C (59°F) in winter to 22°C (~71.5 °F) during the summer months.

Average annual precipitation is around 495 mm (19.5 in), with most rainfall occurring between November and April, while little or no precipitation falls during the summer months. There is an annual average of 70 days with measurable precipitation. Summers in the Monterey Bay area are generally cool and foggy.

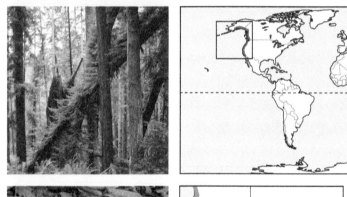

SANTA CRUZ

CALIFORNIA, USA

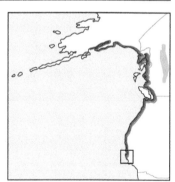

The banana slug can be found on the floors of the Pacific coastal rainforest belt from southeastern Alaska to Santa Cruz, California, where the project is located. Banana slugs are found in relatively mild and humid climates due to their thin, permeable skin. Due to the need to remain moist at all times, banana slug activity is primarily nocturnal. (Photographs courtesy of: habitat, P. Guerrero; banana slug, D. Siciliano)

Animal Physiological, Behavioral & Anatomical Elements
Slugs are most active after rain because of the wet ground, while during dry summer conditions they cover themselves in damp areas under fallen logs or rocks. Breathing occurs through a pallial lung — a heavily vascularized tissue of the dorsal body wall, or mantle, that allows for gas exchange. The pneumostome is an opening in the mantle that lets air into the mantle cavity/lung. The rate at which it opens and closes relates to external temperatures, metabolic rates and hydration levels. Slugs also breathe through their skin, which allows for passive gas exchange between the slug and the atmosphere. On the head of the slug are four tentacles used to sense its surroundings: the two upper optic tentacles observe light and movement, while the two lower sensory tentacles discern chemicals. The slug's mouth is located at the base of the head, where its file-like tongue is used to scrape up food.

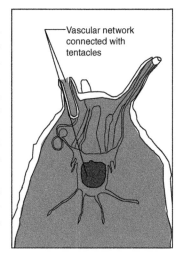

Vascular network connected with tentacles

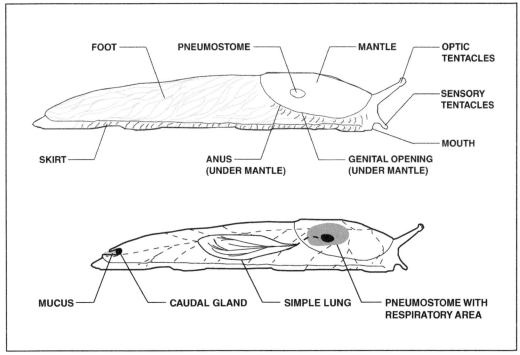

The banana slug contains a pneumostome to allow for gas exchange. Inside the mantle cavity the slug has a highly vascularized section of tissue that facilitates gas exchange. The vascular network diagram is adapted from a diagram of a snail, assumed to be similar. (Diagram adapted from T.R. Jones; general anatomy diagram adapted from Wikimedia & R.A. Barr [1927]; project team; A. Bang, M. Alam: & J.S. Pedersen)

Interface between the Skin & External World Banana slugs live in humid environments to ensure moist skin. They also secrete mucus to prevent desiccation, aid in respiration across a moist surface, find and attract mates, and aid in movement, among other functions. Slime deters predators by its consistency but also because it has anesthetic, or numbing, properties.

During periods of unfavorable weather, such as hot spells or cold temperatures, the banana slug estivates, covering itself under leaf litter and secreting a layer of mucus around its body. Banana slugs are named for their characteristic yellow coloration. They often have golden yellow bodies with dark spots; however, they do exhibit variation in color and can be brown, black and sometimes white. These color patterns can vary with changes in diet, moisture, light availability, age and health.

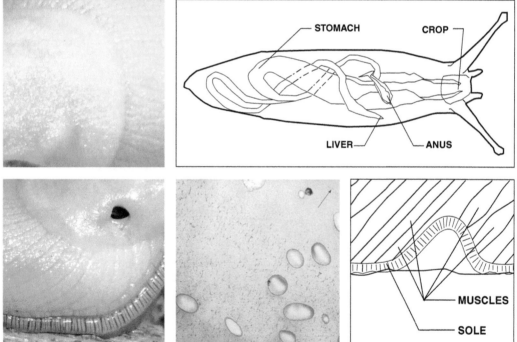

The foot is the largest part of the body, consisting of many tiny muscles that help the slug crawl about in wavelike motion. A microscopic view of banana slug mucus granules; these granules are broken open and mucus is released. (Photographs courtesy of: slug close-ups, D. Siciliano; internal anatomy diagram adapted from G.M. Barker [1999]; mucus granules, Luehtel et al. [1991])

Interrelationship between the Skin & Internal Systems   Slugs produce multiple types of mucus, ranging from thin and watery to thick and gooey, which serve different functions. Thick mucus protects the slug's body from the surface of the ground, prevents dessication, and helps in locomotion. Together with the foot's muscular contractions, this type of mucus helps the slug move forward. Another layer of mucus is thin and watery and is spread from the center of the foot to the edges of the foot. The mucus is made up of fibers that prevent the slug from slipping down vertical surfaces. It also has thixotropic qualities, meaning that when the slug is standing still, the mucus sets into a comparatively firm mass, but as soon it starts moving the mucus liquefies. Mucus is hygroscopic, which means that it rapidly attracts water molecules from the surrounding environment. Dry mucus is contained in granules, and when broken open mucus is released as necessary.

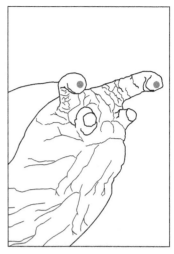

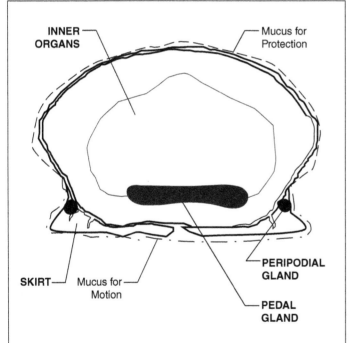

The protective mucus is secreted from the top of the slugs back and then distributed over the body. The mucus that helps the slug move forward is secreted from tiny glands in front of the foot and acts as a track that the slug can glide along. (Photographs courtesy of: standing, D. Siciliano; mucus trail, P. Guerrero; diagram adapted from R.A. Barr [1927])

Proto-Architectural Project   The program of the building is a greenhouse placed in humid Santa Cruz where people can come to observe and learn about forest ecosystems. Four characteristics of the slug were initially analyzed and explored in relation to the design of building: 1) the skin's porosity and permeability which enable breathing; 2) the mucus' ability to protect the slug against both dessication and predators; 3) the slug's ability to adapt to changes in humidity and temperature in relation to the surrounding environment; 4) the slug's ability to maintain balance and communication with its environment through homeostasis. The greenhouse has an adaptive envelope which adjusts and changes according to weather conditions. The envelope fluctuates, changing the internal environment by allowing sun, air and rainwater to permeate inside when desired, and shielding the vegetation from the elements in other instances.

The greenhouse has both an irregular elliptical shape and foundations, thus adapting to the irregularity of the tree canopies above and to the organic uneven forest floor. The process of rainwater collection — later used for plant irrigation – changes the performance, look and feel of the envelope.

Project Documentation The structure is a steel grid which holds the aggregation of bladders, or cushions, in place. The majority of the cushions are composed of a two-layered silicone structure filled with water, whereas the roof cushions are only one layer to collect rainwater, exposed to the outside and allowing the rainwater to drip straight into the structure, enabling transport through the envelope. This transport mainly happens through a secondary structure of woven plastic tubes within the cushions. As the water fluctuates the bladders inflate or deflate, forcing the envelope to close or open accordingly. During rainfall, the silicone units fill, get heavier, and start to droop downwards and open up the wall as they tug the clamps holding them in place. This allows rain to flow into the greenhouse to irrigate the plants, while the overflow is stored for further irrigation. As the envelope dries up and the water dissipates, it will close off the wall.

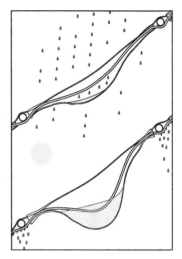

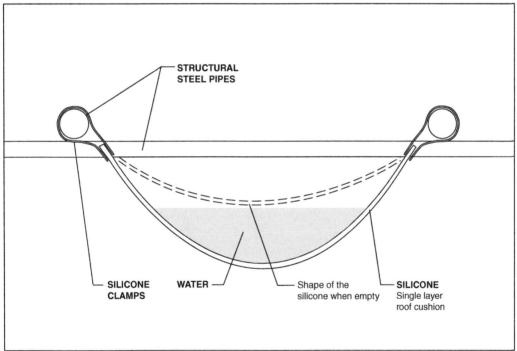

Roof cushion detail open to collect rainwater. The envelope uses the cushion deformations provoked by the water to close off or open up according to the moisture in the atmosphere just as the slug uses its mucus to regulate the degree of air and water exchange through its skin.

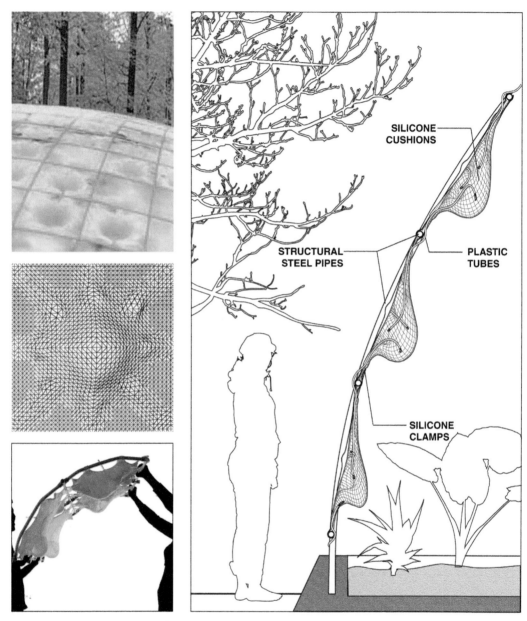

The wall section provides details of the integrated systems within the envelope. The clamps that fasten the silicone cushions to the structure are made of a thicker, less elastic silicone, thus making the clamps substantially rigid. The cushions themselves can stretch and expand according to the quantity of the contained rainwater. The photograph shows a mock-up testing four cushions.

The interior of the greenhouse is dominated by droopy water-filled silicone cushions. This interior volume changes over the day and year. The envelope is active and flexible, registering the ever changing climatic and environmental conditions.

Animal Examples Amphibians are unique among terrestrial vertebrates in possessing skin that is permeable to both air and water. Even though many amphibians are primarily terrestrial, they live closely associated with water in moist environments. The black-bellied slender salamander is in a family of salamanders that lack lungs; that means they only respire through their skin and the tissue inside their mouths. They are most active at high humidity levels and are found underground or under rocks and logs when the humidity levels are low. The squirrel treefrog is also found in moist habitats and hides under bark and leaves when inactive. Eggs are laid and larvae develop in ponds and swamps. The granular dart frog has bright warning coloration, and is found in lowland tropical forests. Adults are ground dwelling, but after the eggs are laid and develop into tadpoles, the mother brings them to water-filled bromeliad plants in the tree canopy.

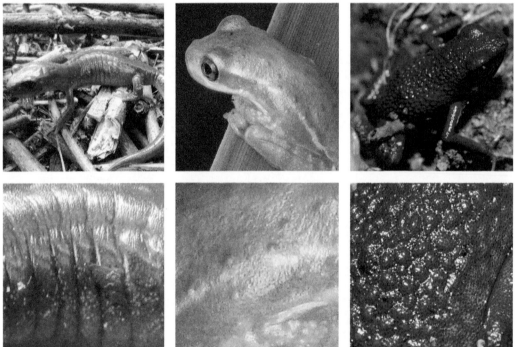

Amphibians primarily obtain water from their environment through passive diffusion through their skin. (Photographs courtesy of: salamander, M. Hedin; lungless salamander & squirrel treefrog, S. McCann; granular dart frog, J. Rothmeyer)

Dyeing dart frog
Dendrobates tinctorius

Phylum:	Chordata	Family:	Dendrobatidae
Class:	Amphibia	Genus:	*Dendrobates*
Order:	Anura	Species:	*D. tinctorius*

Photograph courtesy of M. Bartelds

Habitat & Climate   Poison dart frogs are native to moist environments in Central and South America and are all members of the Dentrobatidae family. Many have colorful skin patterning and are toxic. *Dendrobates tinctorius*, or the dyeing dart frog, occurs in the eastern Guiana Shield, in southeastern Guiana, Suriname, French Guiana and northeastern Brazil. They occur in isolated populations in the uplands of the Guiana Shield, preferring elevations of 200 to 600 m (~650–2000 ft). This species does not prefer lowland forests, perhaps because they experience seasonal flooding, and the frog is primarily ground dwelling. With an average of 373 cm (147 in.) of rain per year, precipitation is highest between April and June, and lowest between August and September. Yearly relative humidity averages 76.5% during the summer months, with peak levels in May (83%).

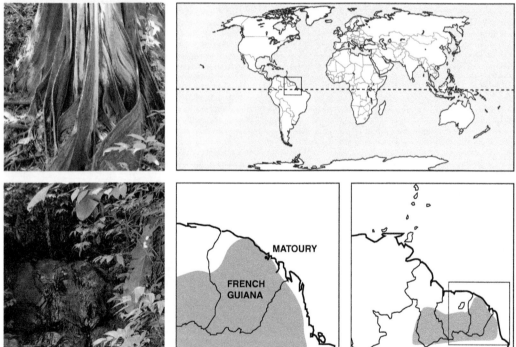

The forests of French Guiana remain moist through much of the year, providing ideal habitat for the dyeing dart frog. *D. tinctorius* is primarily ground dwelling and can be found in tree roots, on vines or under rocks, where it forages in the leaf litter. (Photographs courtesy of M. Bartelds)

Animal Physiological, Behavioral & Anatomical Elements
Like other poison dart frogs, *D. tinctorius* occurs in extremely
humid areas, most often near running water. They are most active
during the rainy season, foraging in canopy-gaps on arthropods.
In the dry season they retreat to protected areas in trees and
plants and consume less prey. They are diurnal, meaning they are
active during the day, particularly in the early morning and late
afternoon.

Unlike many other vertebrates, frogs do not have waterproof skin,
and their lifestyle is very closely associated with water. In fact,
larval development takes place in the water. While lungs function
in gas exchange, frogs breathe primarily through their skin. Their
skin has to remain moist for breathing, and this is done with the
aid of mucus glands in their skin.

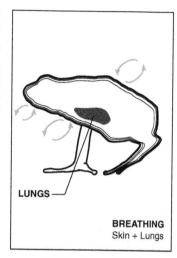

LUNGS

BREATHING
Skin + Lungs

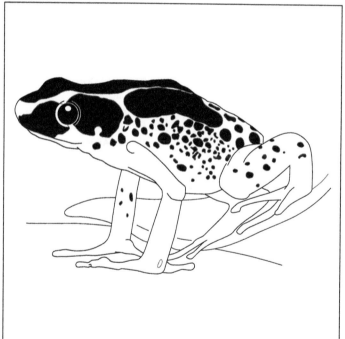

The skin aids in thermal regulation, UV protection, warning predators, respiration and toxicity. The pigment melanin
absorbs heat during the day and helps block UV light. Each foot has four toes which possess a pad that functions as
a suction cup used for gripping. (Photograph courtesy of M. Bartelds; project team: E. Lani & J. Su)

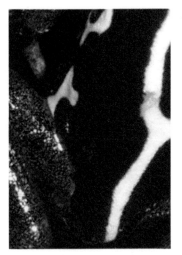

Interface between the Skin & External World Frog skin is permeable and moist. Their skin consists of two main layers, the epidermis, or outer layer, and the dermis, the inner layer. Poison dart frogs are recognized for their bright coloration and toxic skin. The color patterns serve as a warning to predators; they are a signal of their toxicity and distastefulness. The skin contains specialized glands located in the dermis that produce poison. The poisons are neurotoxins, called batrachotoxin, which act on the nervous system of the animal that ingests them. Poison is squeezed out of the glands when muscle cells around the glands contract. Toxicity is thought to be acquired from the diet, because in captivity frogs often lose toxicity. Mucus glands are much smaller and produce mucus to keep the skin moist. This moisture helps prevent desiccation and also serves gas exchange functions. Both types of glands are invaginations of the epidermis.

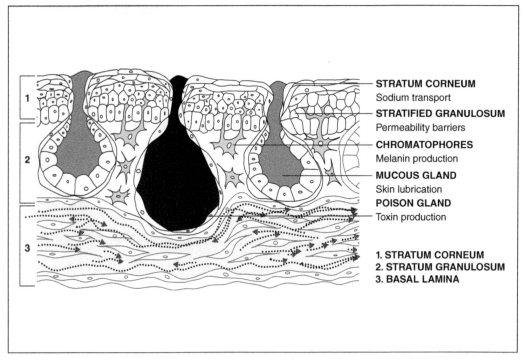

	STRATUM CORNEUM
	Sodium transport
1	STRATIFIED GRANULOSUM
	Permeability barriers
	CHROMATOPHORES
	Melanin production
2	MUCOUS GLAND
	Skin lubrication
	POISON GLAND
	Toxin production
3	1. STRATUM CORNEUM
	2. STRATUM GRANULOSUM
	3. BASAL LAMINA

Skin coloration patterns are produced from pigmented cells, called chromatophores, which are located in the dermis. Because dyeing dart frogs occur on isolated islands, each population possesses striking variation in its skin coloration patterns. *D. tinctorius* can be any combination of blue, yellow, white and black with stripes and spots. (Photograph courtesy of M. Bartelds)

Interrelationship between the Skin & Internal Systems Gas exchange occurs through the lungs and the skin. Oxygen passes directly through the membranous skin and into an extensive network of blood vessels. When using their saclike lungs, frogs breathe through opening their mouth or nostrils and allowing air to flow through. Once oxygen makes it into the blood stream, the oxygen-enriched blood enters into the heart, which is made up of only three chambers, two upper atria (right and left) and one lower ventricle. Oxygen-enriched blood is forced into the arteries and disperses to tissues throughout the body, while oxygen-poor blood is enriched through the skin and lungs and then distributed accordingly. The drink patch, densely populated capillary tissue located on the underside of frogs, can be submerged or touched against a moist surface to hydrate the body.

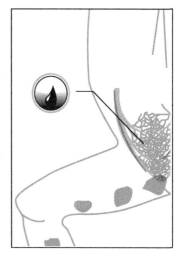

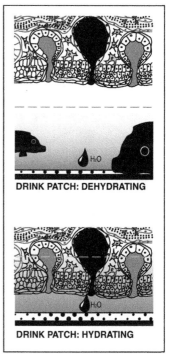

DRINK PATCH: DEHYDRATING

DRINK PATCH: HYDRATING

The drink patch is located between the hind legs on the lower belly, enabling the frog to easily lower it into the water to rehydrate. The frog simply dips its pelvic area below the water surface or it presses the area against a moist surface. Water molecules are wicked through the skin by osmotic action and delivered to organs throughout its body on the cellular level.

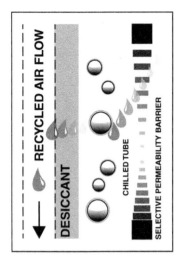

RECYCLED AIR FLOW

DESICCANT

CHILLED TUBE

SELECTIVE PERMEABILITY BARRIER

Proto-Architectural Project   The museum, located in Matoury, is dedicated to Amazonian cultural history. It houses both indigenous and imperial artifacts and art. Lifted off the ground, the building is designed with a concentric plan, making humidity and thermal regulation manageable. The more moisture tolerant artifacts are located closer to the perimeter while the more delicate objects are placed in the center. Using the *D. tintorius* as its inspiration, this project has led to an architecture that mediates humidity levels from an exterior expanse to interior, intimate spaces. Density, layering and curved surfaces play a large role in its aesthetic, but most specially it functions to dehumidify the museum space. Rather than capturing water that is generously available in the rainforest, it explores how to keep humidity out. The humidity regulation is mostly passive, but supplemental mechanical systems are included for use in extreme weather.

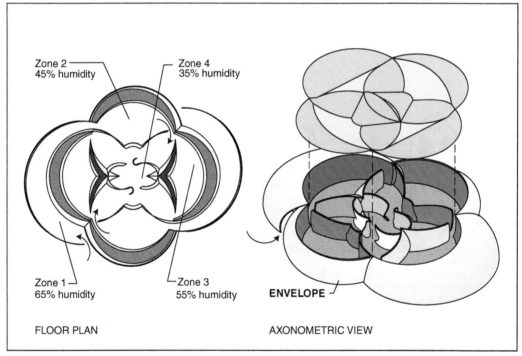

Zone 2
45% humidity

Zone 4
35% humidity

Zone 1
65% humidity

Zone 3
55% humidity

ENVELOPE

FLOOR PLAN

AXONOMETRIC VIEW

The design seeks to wick water out of a space by capturing it in a desiccant-like structural system. Walls become denser and thicker as they approach the center of the symmetrical plan. This removes humidity at the graded percentages and surrounds the most precious artifacts with a large buffer zone, intensifying the experience of depth and layering within the walls.

Project Documentation   The first layer is the permeable one, a hard material with more or less perforation according to the specific humidity desired. The second layer is a paneled network of tubes filled with chilled water. The cooler temperatures within the tube attract water through condensation, which is absorbed by the third layer, a structured foam desiccant. The desiccant maintains its shape and cannot degrade too readily under moist conditions, with a cellular anatomy creating space between molecules large enough for water to pass through. Behind this is a second "selectively" permeable layer, 6, that acts as a surface weep screed, allowing water to weep from the desiccant into the wall cavity, but preventing the re-admittance of water into layer 3. Through the main structure and a secondary tension structure, dry air would flow wicked water to a designated catch, to be dumped under the waterproof "belly" of the building, to act as a cooling mechanism.

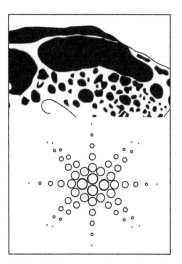

Top diagram shows how the selectively permeable layer is inspired by the scalar coloration of *D. tinctorius'* skin. The layering of the various wall tiles is necessary to dehumidify but also to create a depth and scale that visitors perceive as their museum visit progresses.

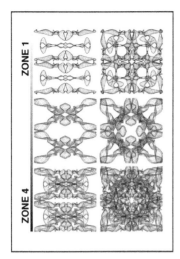

ZONE 1

ZONE 4

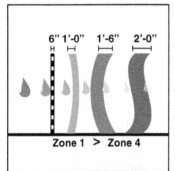

6" 1'-0"   1'-6"   2'-0"

Zone 1 > Zone 4

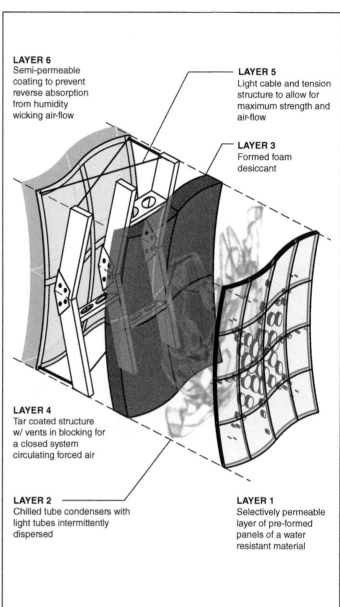

LAYER 6
Semi-permeable
coating to prevent
reverse absorption
from humidity
wicking air-flow

LAYER 5
Light cable and tension
structure to allow for
maximum strength and
air-flow

LAYER 3
Formed foam
desiccant

LAYER 4
Tar coated structure
w/ vents in blocking for
a closed system
circulating forced air

LAYER 2
Chilled tube condensers with
light tubes intermittently
dispersed

LAYER 1
Selectively permeable
layer of pre-formed
panels of a water
resistant material

Exploded wall section illustrating each layer: 1) selective permeable layer; 2) chilled condensers and lighting; 3) structured foam desiccant; 4 & 5) lightweight post & tension structure; 6) secondary selectively permeable layer.

The museum spaces surround visitors with their walls composed of capillary-like tiles that facilitate air dehumidification. Illuminating the room are lights with blue hues, while highlighting artifacts and walls are yellow hues.

Animal Examples Marine invertebrate animals, especially those that live in intertidal zones, have to deal with the pressures of desiccation as well as variations of water salinity. Chitons inhabit shallow waters and have eight hard plates covering their bodies. For protection from these environmental pressures they often live under rocks or in rock crevices. They quickly regulate water balance in their cells when salt levels are too high or low. Blood sea stars are found in coastal regions and can be found on or under rocks to protect themselves from predators and external elements. Barnacles are sessile crustaceans that primarily live in the intertidal zone. They have shells that help prevent them from desiccation. The strawberry anemone is another sessile animal. However, they can detach themselves from the substrate when conditions such as desiccation cause them to move to another location. They regulate the salt concentration in their tissues through diffusion.

Many marine invertebrates live in harsh environments where they face large variations in water availability and salinity. (Photographs courtesy of: chiton, R. Muthukrishnan; blood sea star, A. Jaffe; barnacles, morgueFile.com; strawberry anemone, M. Bartosek)

Ochre sea star
Pisaster ochraceus

Phylum:	Echinodermata	Family:	Asteriidae
Class:	Asteroidea	Genus:	*Pisaster*
Order:	Forcipulatida	Species:	*P. ochraceus*

Photograph courtesy of H. Thomssen

Habitat & Climate *Pisaster ochraceus* inhabits rocky shores, kelp forests and estuaries along the Pacific coast of North America. It is most common in the rocky intertidal zone, one of the harshest and most dynamic environments on Earth. Rocky intertidal organisms are subjected to myriad predators from both the land and sea, as well as intense and nearly constant wave action. In addition, they are exposed to air and direct sunlight up to two times per day at low tide. While *P. ochraceus* has an extensive range along the Pacific coast, this project focuses on southern California, with a Mediterranean-type climate of sunny, warm summers and mild, wet winters. The rainy season (November to March) rarely produces more than 30 cm (12 in.) of rainfall. Ocean temperatures range from 13 to 21°C (56–70°F), with an average relative humidity of ~65%. Temperatures rarely exceed 85°F (29.5°C) in summer, nor drop below 50°F (10°C) in winter.

Pisaster ochraceus is abundant on rocky, intertidal shores along the entire Pacific coast of North America from Alaska to Baja, California. Animals that live in these harsh habitats must be adapted to survive intense wave action and to resist desiccation at low tide. This project is focused on the intertidal zone at the Santa Monica pier. (Photographs courtesy of: beach and tide pool, N. Lucas; pier, R. Molina)

Animal Physiological, Behavioral & Anatomical Elements

Pisaster ochraceus can grow to a diameter of 25 cm (10 in.) with five arms radiating from the center. It is found in two distinct color morphs, with purple being slightly more common than orange. It has no centralized nervous system (i.e., brain and spinal cord), but instead a series of connected nerves that run throughout its body.

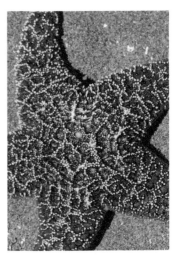

A unique physiological trait of sea stars and other echinoderms is a water vascular system. The system uses a network of canals and muscles to move water to transport food and waste and for respiration. It also uses the system to create and relieve suction among hundreds of tube feet independently, allowing the animal to "walk" along the substrate. This means that *P. ochraceus* can readily move around underwater but must remain largely in place while exposed at low tide.

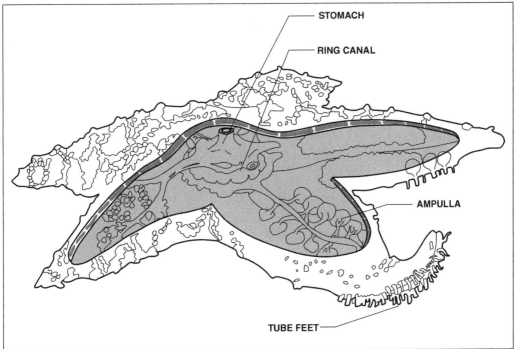

The figure above illustrates the internal pneumatic ring canal (circulatory), stomach (digestion), ampulla (circulatory), and tube feet (locomotion/predation). The ring canal moves water taken in from a dorsal valve (madreporite) and channels it through the arms of the ochre sea star to the ampulla and finally to the tube feet as it moves in the water. (Photograph courtesy of D. McShaffrey; project team: P. Mecomber & A. Ariosa)

164

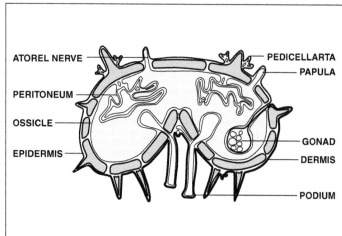

ATOREL NERVE

PERITONEUM

OSSICLE

EPIDERMIS

PEDICELLARTA

PAPULA

GONAD

DERMIS

PODIUM

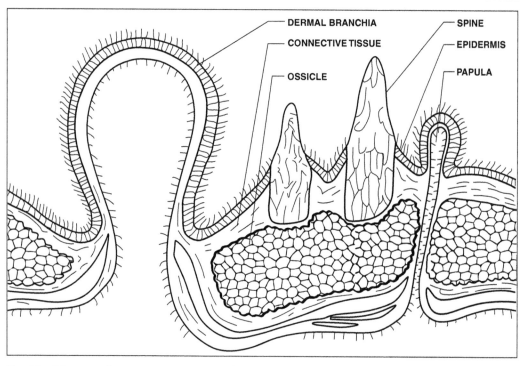

DERMAL BRANCHIA

CONNECTIVE TISSUE

OSSICLE

SPINE

EPIDERMIS

PAPULA

The skin of *Pisaster ochraceus* is a continuous epidermis with embedded bony plates called ossicles. The ossicles create a network of armor plates that are locked closely together when out of the water, producing a hard and tough skin. A strong vacuum is also produced inside its many tiny tube feet, making it nearly impossible for waves or predators to dislodge the animals from their rocky substrate. Evaporation of water through dermal branchiae helps keep the animal cool while exposed to sunlight and air. (Photograph courtesy of D. McShaffrey; drawing adapted from R. Fox, Lander University)

Interrelationship between the Skin & Internal Systems   Living in the intertidal zone, *P. ochraceus* must be adept at handling two entirely disparate environments. Its body is essentially a pneumatic machine pumping water through its water vascular system to move its limbs. The system branches out to five radial canals, one for each arm of the sea star.

Desiccation, the loss of water via evaporation, is critical in setting the upper limits of many intertidal species. However, *P. ochraceus* can tolerate a loss of up to thirty percent of its weight in bodily fluids. It also behaviorally combats desiccation by hiding under rocks to avoid direct sunlight, but will move about as it searches for food — particularly for the California mussel, *Mytilus californianus*, an abundant intertidal organism that has its own unique adaptations for dealing with the harsh habitat.

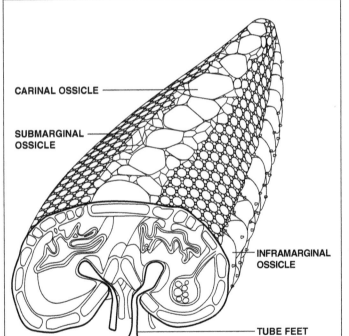

CARINAL OSSICLE

SUBMARGINAL OSSICLE

INFRAMARGINAL OSSICLE

TUBE FEET

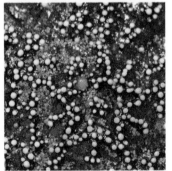

When underwater, the ochre sea star becomes softer and more malleable, as soft tissue is exposed to aid in respiration and waste excretion, and water is pumped in and out of the tube feet for locomotion. The skin is also covered with tiny pinchers called pedicellaria that are used underwater to prevent larvae of other marine invertebrates from settling and growing on its surface. The ability of *P. ochraceus* to respond to its dynamic intertidal environment in these ways is largely made possible by the water vascular system described above. (Photographs courtesy of D. McShaffrey; drawing adapted from R. Fox, Lander University.)

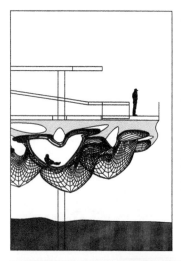

Proto-Architectural Project The project is positioned in the intertidal region of Santa Monica, California, a natural habitat for the ochre sea star. The iconic Santa Monica pier, jutting out hundreds of feet over the ocean, is the perfect place for tourists to experience this region. The proposal re-interprets both the Santa Monica pier and the skin of *P. ochraceus*. The project is designed under the pier, using the existing infrastructure to place a new spa and observation platform. To enter, one descends beneath the pier, into the underbelly of the structure through a series of connecting pods. A hard shell holds a series of expandable pods containing portions of the spa that expand and submerge underwater as the tide rises. Once inflated, they become stable environments, creating a habitable and desirable space in a previously inaccessible and dangerous place. It is an escape into an unfamiliar environment of undersea creatures.

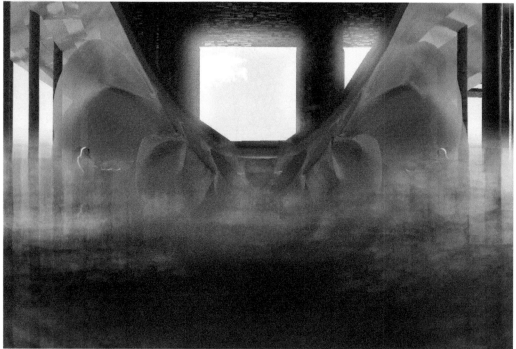

View of the spaces beneath the pier. Each pod is inhabitable by one or a few individuals and offers a unique perspective of the environment surrounding the viewer. Depending on the tidal level, the viewer is above water level or submerged, co-existing with the ocean environment.

Project Documentation   The project is inspired by the sea star's skin system, which is not entirely solid. Between hard and impervious panels, there is a series of soft membranes that pneumatically expand as the tide rises and interfaces with the assemblage. These membranes inflate and become spaces under water, carving out new real estate and creating a temporarily dynamic interface between humans and an environment typically unsuitable for them. The walls of this membrane vary in translucency, allowing views of the environment. The temperature of these spaces is consistently cooler than the external temperatures, ideal for the peak tourist season during the summer. The membrane is firm to the touch, yet appears soft. The material composition of the membrane is a series of layers, the outermost of which is similar to ETFE. In between there are layers of sponges that when submerged absorb water and displace air.

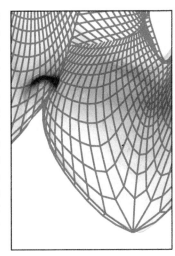

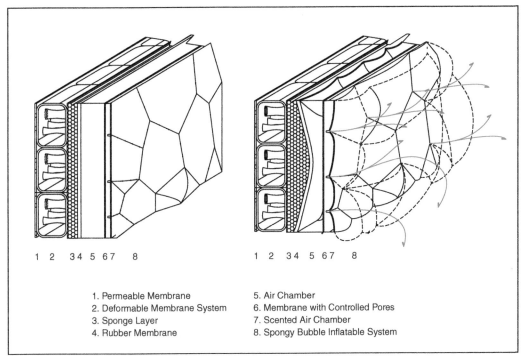

1 2   34 5 67   8   1   2   34 5 67   8

1. Permeable Membrane
2. Deformable Membrane System
3. Sponge Layer
4. Rubber Membrane

5. Air Chamber
6. Membrane with Controlled Pores
7. Scented Air Chamber
8. Spongy Bubble Inflatable System

Here a wall section detail depicts the intricacy of the envelope composition. The intent of the architectural envelope is to provide maximum sensory experience. The soft membrane is composed of multiple layers capable of exchanging volumes of water and gasses with the surroundings, varying the wall thickness when above or below water.

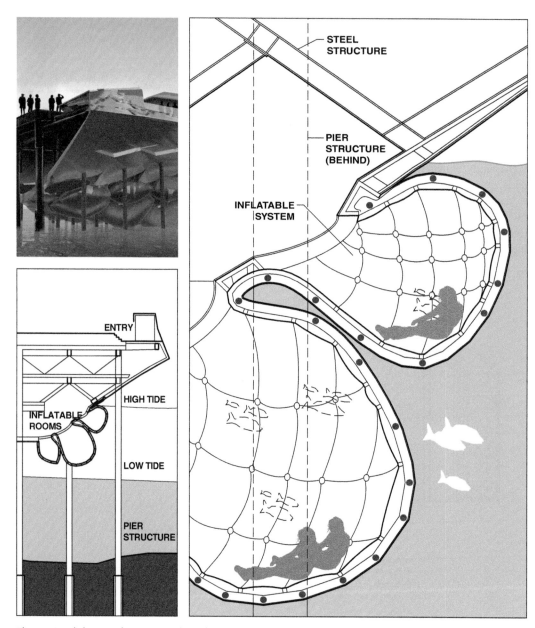

The sectional drawing demonstrates how the system hangs, anchored to the underbelly of the pier. It is not unlike the sea star's capability to cling to the rocky shores of the coastline. The pillowed spaces are dynamic environments with no defined floor, no wall, no ceiling, but rather an encompassing padded hollow space which can be utilized in myriad ways from standing, to sitting, to sprawling. All of these details mimic the experience and function of the sea star abstracted in the form of a multi-sensory environment.

As a contrast to the amusement park environment above, the space of the intertidal zone opens up real novelty — the novelty of experiencing an environment not designed for the user, where architecture protects and allows for impossible interactions. This becomes a region of water and mist, of dappled light and shadow, alien creatures, and strange interactions. It is a harsh environment full of perils to its inhabitants. The project creates a calm space within this tumultuous environment.

Animal Examples Gila monsters are large, desert-dwelling lizards that spend almost all of their time in underground burrows, rarely coming out to feed. These traits allow gila monsters to prevent water loss in the arid habitats in which they live. Horned lizards have evolved a unique method of collecting water during desert rains. They possess narrow channels in between their scales that cause water to flow, through capillary action, into their mouths. Kangaroo rats live in burrows underground that they cover with dirt during the day to prevent water loss. While they sleep kangaroo rats bury their noses into their fur to recycle water as they breathe. Honeypot ants construct nests far underground where the soil is cool and moist. Some ants in each colony act as specialized storage devices; their abdomens are able to swell to the size of marbles in order to store nectar, which provides nutrients and water to other individuals in the colony.

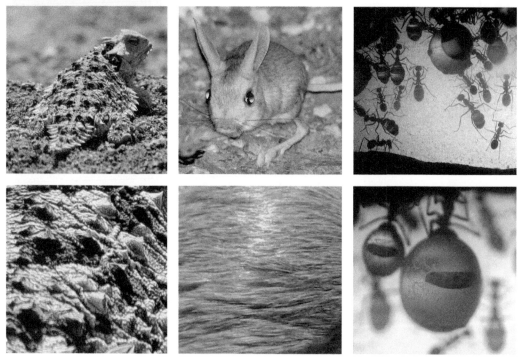

Desert-dwelling animals cope with low availability of water using adaptations that help them acquire water and prevent them from losing water. (Photographs courtesy of: gila monster, Wikimedia; horned lizard, J. Rothmeyer; kangaroo rat, S. Yeliseev; honeypot ants, T. Doyle)

Namib Desert beetles
Onymacris unguicularis, Physasterna cribripes

Phylum:	Arthropoda	**Family:**	Tenebrionidae
Class:	Insecta	**Genus:**	*Onymacris, Physasterna*
Order:	Coleoptera	**Species:**	*O. unguicularis,*
			P. cribripes

Photograph courtesy of A. Sosio

Habitat & Climate   The Namib Desert, situated on the western coast of Africa in Angola, Namibia, and South Africa, is the world's oldest and one of its harshest deserts. The desert is broadly characterized into two regions, the coast and inland. The temperature regime varies between the coast from 9–20°C (48–68°F), and inland from 0–45°C (32–113°F). Precipitation in each region comes from rainfall, but in the coastal region it also comes in the form of dense, enveloping fog. While the inland region receives about 5 mm (0.02 in.) of precipitation per year, the coastal region can get up to 200 mm (7.8 in.) in the wettest areas. Rain and fog are the only opportunities for desert life to acquire water since almost no bodies of water are located in the Namib Desert. The diversity of animal life is composed mainly of arthropods, notably darkling beetles, commonly referred to as Namib Desert beetles, adapted to hot and dry desert conditions.

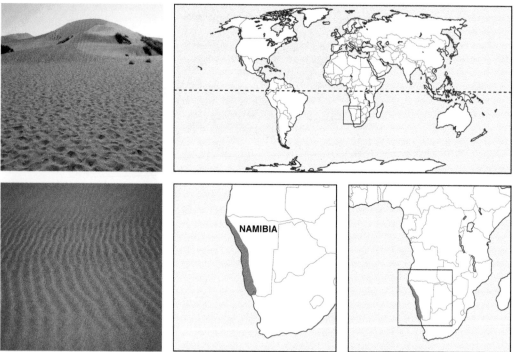

The project is set in the coastal region of the Namib Desert in the country of Namibia. This area is characterized by vast, undulating sand dunes colored pink and orange. Vegetation is sparse and consists of lichens and succulent plants near the coast and shrubby vegetation farther inland. (Photographs courtesy of A. Sosio)

Animal Physiological, Behavioral & Anatomical Elements
Due to an extreme scarcity of water, some desert beetles have
evolved strategies to harvest water from the environment. The fog
that arrives in late evening and early dawn represents a more reli-
able source of water than rain, so beetles have ways to collect
fog droplets. These include sucking water from sand covered in
fog, drinking from water that has collected on vegetation, creating trenches in the sand to collect fog, and finally, fog basking.
Fog basking is a behavioral strategy that involves collecting water
using body surfaces. The beetles face into the wind and assume
a tilted position with their heads close to ground, permitting fog
droplets to accumulate on their backs and trickle down the body
into their mouths. Additionally, they often have long legs to help
with movement across and within sand as well as to raise them
off the hot substrate.

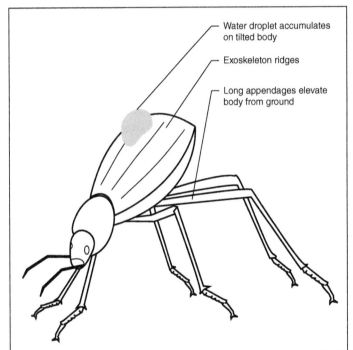

Water droplet accumulates
on tilted body

Exoskeleton ridges

Long appendages elevate
body from ground

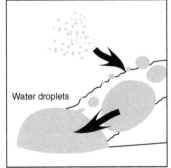

Water droplets

Two species of Namib Desert beetles, *Onymacris unguicularis* and *Physasterna cribripes*, are nocturnal, burrowing
deep under the sand during the day and emerging at night to feed. Physiological adaptations include excreting very
little water in their waste products and having a low metabolic rate. Many of these beetles also have a waxy surface
over their exoskeletons to prevent the loss of water. (Photograph courtesy of Nørgard and Dacke [2005]; project
team: E. Chen & C. Rodriguez)

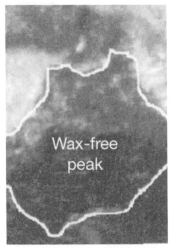

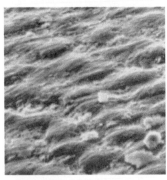

Interface between the Skin & External World Two species are known to fog bask: *O. unguicularis* and *O. bicolor.* Previous studies have examined the mechanism by which fog is collected. There are regions on the elytra surfaces that are coated in wax (troughs), making them hydrophobic (repellant to water), and regions that are not covered in wax (peaks), which confer hydrophilic (water-attracting) properties. The small fog droplets congregate on the hydrophilic areas until they form large enough drops, which funnel down to the mouth. Numerous biomimicry studies have been based on this water collection mechanism; however, they were incorrectly based on a species that does not fog bask. Nevertheless, the physical properties of the hydrophilic peaks that attract fog droplets have been very successful in designing prototypes and products. This chapter studies *O. unguicularis,* known for collecting fog, and *P. cribripes* for its troughs and peaks.

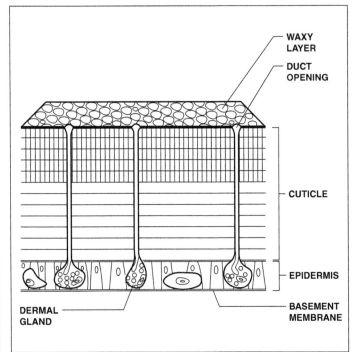

The surface of *P. cribripes* with peaks and troughs was used as the inspiration for the project. The bumpy surface of the elytra is shown with the wax-free, hydrophilic and waxy, hydrophobic areas. Though not biologically accurate, the physical properties of the water collection mechanism can inspire design. A cross section of the beetle exoskeleton showing the waxy layer. (Photographs courtesy of: bump stain and micrograph, Parker and Lawrence [2001]; dorsal view, Nørgard and Dacke [2005])

Interrelationship between the Skin & Internal Systems   Insects breathe using a tracheal system, a series of branching tubes that connect to the external environment through openings called spiracles. Tracheae branch into smaller tubes called tracheoles that conduct oxygen directly to every cell in the body. During respiration oxygen is gained but water is also lost through evaporation. In the desert any lost water comes with a huge cost. Therefore, their respiratory systems have evolved to prevent water loss. Namib Desert beetles have lost the ability to fly; their outer wings, elytra, have been fused together to form a hardened surface with a cavity underneath. The beetles breathe into the cavities, which helps reduce respiratory water loss. During the day they burrow into sand to further prevent water loss. Finally, they have long periods where they stop breathing to minimize the loss of water.

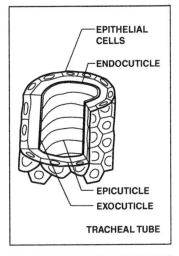

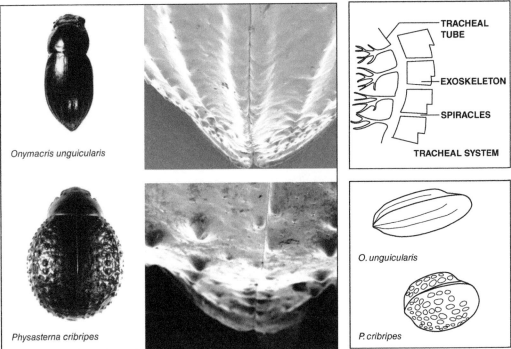

The efficiency in water retention varies among the different beetle species. This project takes two of the most effective beetles and combines the geometry in order to optimize water retention capabilities. The first is *O. unguicularis,* in which we mimic the ridges and valleys of the shell, and the second is *P. cribripes,* which is the inspiration for the bumpy texture. (Photograph courtesy of Nørgard and Dacke [2005])

Proto-Architectural Project   The project, located in the Namib Desert, serves as a temporary residence for desert researchers. Taking cues from the beetles' ability to collect water from such an arid climate, the building is positioned so that the largest sloped roof/façade is facing southeast toward the morning fog that sweeps up from the ocean, and burrows a portion of the building underground to regulate its thermal mass.

The massing takes after a combination of two different species' exoskeleton structure in order to optimize the water retention capability. The bumpy texture is inspired by *P. cribripes* as a way to capture the moisture in the air, while the overall geometry was extracted from *O. unguicularis* with the ridges and valleys to channel the water into storage.

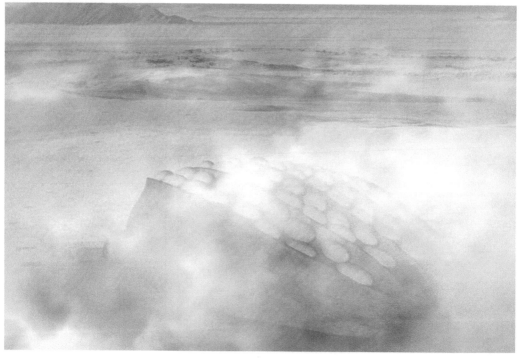

The project's geometry allows for a large portion of the façade to face the direction of the morning fog, with an angle designed so that the water can travel down from the top toward the water storage. Utilizing the characteristics of the beetles' bumps and ridges, the highly articulated building's surface has the capacity to capture water. The water can then be used to create a microclimate for its inhabitant.

Project Documentation   The building envelope uses a series of mesh discs to capture moisture from the morning fog. Each disc is set on a pivot and and tilts when enough water has accumulated. The water rolls down the grooved roof membrane toward the water cistern at the base of the structure. The collected water not only provides a potable water source for the occupants but is a cooling mass to help regulate comfort within the space. Not only does the shape of the building mimic the "fog basking" water collection feature of the beetle, architecturally the form creates an interior volume (low to high ceilings) which optimizes air stratification and overall climate comfort. A portion of the roof discs are perforated to allow for the collection of water to sustain the interior plants. The plants allow for storage of water, food, aromatherapy, and help humidify the space through evapotranspiration.

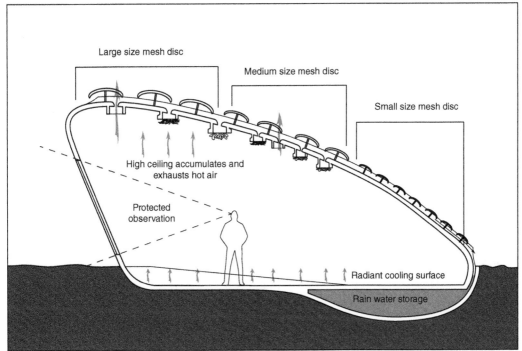

The diagrams show the components of the mesh disc and how it collects water droplets. The schematic building section indicates the exterior positioning and decreasing sizes of the discs as well as delineating the interior space, slightly sunk into the sand to take advantage of its thermal mass.

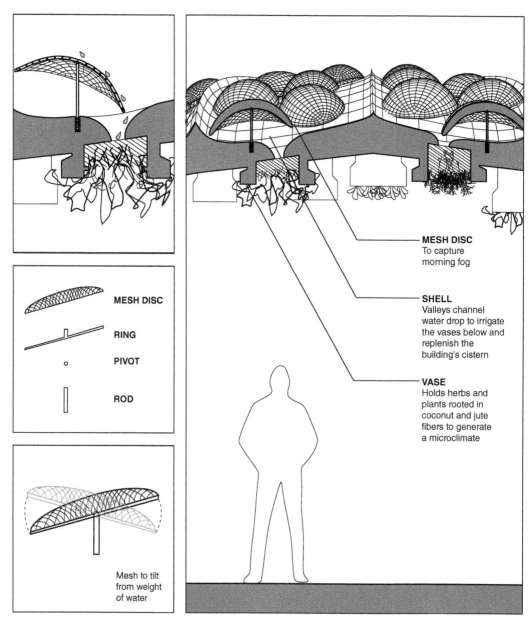

MESH DISC
To capture
morning fog

SHELL
Valleys channel
water drop to irrigate
the vases below and
replenish the
building's cistern

VASE
Holds herbs and
plants rooted in
coconut and jute
fibers to generate
a microclimate

MESH DISC

RING

PIVOT

ROD

Mesh to tilt
from weight
of water

The wall section and rendering demonstrate how water is captured through the discs made out of mesh. The water is then funneled into the channels by the geometry of the overall massing. The channels are connected to the upside down pots in the ceiling which grow a range of mixed herbal vegetation. The vegetation shown on the section creates a ceiling condition that is unexpected for a desert condition and provides a small quantity of edible greens.

In this hot, arid and typically deemed uninhabitable environment, the project provides an unexpected oasis for researchers. Furthermore, the dynamic system on the roof engages in water collection, transforming the interior with landscapes of edible vegetation, thus creating an environment that allows the researcher to thrive in such a severe climate.

6 Protection

The ant species *Cephalotes atratus* protects and defends its colony by using heads of the workers to block its nest entrances. (Photograph courtesy of S. Price)

Many of the skin functions we have previously considered could be interpreted as protective adaptations. For example, animals use communication, often through coloration, to prevent themselves from getting eaten by a predator or attacked by a competitor. This may occur in the form of warning coloration or camouflage. Animals continuously thermoregulate to protect themselves from environments that are too hot or cold. They also use their skin to protect themselves from losing water. In this section, we will consider two additional ways in which animals protect themselves — by using physical defense to protect themselves from attack, and by chemical defense from predators and the environment.

The skin and its appendages provide the first line of defense for most animals, and many have evolved extreme modifications of the skin that function in protection against predators or against members of their own species. These include horns, antlers, sharp teeth, legs, claws, spines and sharp hairs. The skin itself can be hardened to serve as protective armor. The shells of turtles, snails, and clams are examples of hardened animal coverings that prevent predators from getting to the vulnerable, internal body structures. Scales, bony plates and fur are existing appendages born from the skin that are used to protect from the elements and from predators. Animals also use behaviors in conjunction with skin appendages for protection. For example, erection of feathers or fur by tiny muscles in the skin can make an animal appear larger than it really is and scare potential predators or competitors. A vestige of this innate behavior is retained in humans – hairs

(Top) Tent-making bats protect themselves from rain by sleeping under large leaves that they modify into a tent-shaped form.
(Bottom) Many lizards, such as the banded gecko, will undergo the radical process of dropping their tails to escape predation, leaving the predator with a writhing tail and no prey. (Photographs courtesy of: bats, S. McCann; gecko, J. Rothmeyer)

standing up on the back of the neck during a scary movie are our body's attempt to appear intimidating to an imaginary threat. Many animal species move in groups, or herds, as a protective defense. Their coloration patterns may work in tandem to confuse predators.

Toxins, venom and poison are all chemical defenses that often are secreted by glands in the skin or protrusions from the skin. In many animals these secretions are used for scent marking, but in other cases they are modified to aid in defense from predators. For example, skunks are able to forcefully eject noxious secretions from their anal glands when threatened. A lot of amphibian species have poison glands in their skin which, combined with bright warning colorations, deter predators. A lot of invertebrates secrete chemicals from their joints or specialized glands that are irritating, toxic, waxy, or odorous. Invertebrates are often brightly colored to warn predators of their distastefulness. Some marine invertebrates, like octopuses and squid, produce colored pigments that can be ejected to confuse predators and facilitate their escape.

The fundamental idea of shelter is one of a structure that provides protection from predators and the elements. Over time, the type and meaning of human shelters have been added to. Shelters were initially used as protection from the weather, predators and other human populations. However, shelters not only serve the function of protection, they also serve as a tool of communication. A primary signal shelters send is that of power. For example, castles and defensive forts convey power even without actively being used. Today, structural security intended as defense and safety occurs in a few specialized building types such as embassies, jails or courthouses.

Buildings still serve to protect humans from environmental elements, but now they also need to address more severe challenges such as ozone layer depletion, air pollution, acid rain, and an increased severity of weather conditions resulting from climate change.

One critical environmental element is the sun, our primary source of energy. While the sun's radiation can be turned into an unlimited energy source, it can also produce chemical changes detrimental to humans and the Earth's atmosphere.

The trees provide shade and through evapotranspiration mechanisms cool the protected enviroment. The photovoltaic panels supply the energy necessary to activate the misting cooling system. Ecosistema Urbano, Ecoboulevard, Madrid, 2005. (Photograph courtesy of D. Terna)

Three large artificial trees respond to the scarcity of nature in this peripheral city development, while providing climatic comfort in public spaces. Ecosistema Urbano, Ecoboulevard, Madrid, 2005. (Photograph courtesy of D. Terna)

The problems created by the necessity to both shield and harvest the sun's radiation provide unique opportunities in architectural exploration. From the sun we can acquire energy, and through the design of a smart active building envelope, mediate it, filter it, and transform it for internal use and comfort. The use of advanced materials can further enhance envelope performance. For example, it can include a process involving the sun as a catalyst that is used to produce coverings that contain titanium dioxide. The coverings break down inorganic and organic pollutants and also actively participate in cleaning the air while protecting a building façade.

Here we have used two examples to show different ways of taking insights from nature to design protective structures. Inspired by the pangolin's hard scales, the project is a portable structure that accumulates solar energy to be used in emergencies while protecting inhabitants from the sun's UV rays. The sweat glands of hippopotamuses secrete a substance that protects them from UV rays and helps maintain the animal's water balance. The designed building envelope is constituted of tessellated elements that collect water; these are used to create a thermal mass to serve as heat protection and a water reservoir.

Animal Examples — Protection   Animals have a multitude of mechanisms that serve as protection mainly from predators, but also from the elements and individuals of their own species. Meerkats are social animals; when outside of their burrows one will stand sentry to watch out for predators and alert the other individuals. Hedgehogs roll into a ball and have stiff, hollow hairs they use to protect themselves from predators. Skunks are most known for the noxious liquid they spray from anal scent glands to ward off predators. Lionfish protect themselves through their venomous spines. Porcupine fish also have spines; they take in air, and their bodies inflate, thus protruding the spines. Similarly, sea urchins have spines that serve as protection. The flamingo tongue snail is brightly colored due to the live tissue that covers its shell. When attacked the tissue retreats into the shell.

All animals have adaptations that serve as protection from predators, the environment and members of the same species. These include spines, toxins, hardened coverings and social behavior. (Photographs courtesy of: meerkats, I. Mazzoleni; hedgehog, Wikimedia, J. Hempel; skunk, B. Garrett; lionfish, A. Jaffe; porcupine fish & pencil urchin, J. Maragos; flamingo tongue snai, A. Jaffe)

Animal Examples — Protection   Spiders, such as black widows and orb weavers, have venom to subdue prey but also to protect from predators. The armadillo lizard will roll into a ball, hence the common name, by putting its tail in its mouth and protecting itself with thick, sharp scales.

Harpy eagles are top predators; they use their sharp claws to capture prey. Their coloration patterns help them blend into the forest canopy to sneak up on their prey. Pill millipedes roll into a ball when disturbed. Rhinoceros beetles are large beetles with very thick exoskeletons. Males use horns in combat with other males. Many animals, such as butterflies, have eyespots, markings on their bodies that resemble eyes. When flashed they are thought to mimic other animals to deceive predators.

All animals have adaptations that serve as protection from predators, the environment and members of the same species. These include spines, toxins, hardened coverings and social behavior. (Photographs courtesy of: black widow spider & orb weaver spider, M. Hedin; armadillo lizard, P. le FN Mouton; harpy eagle, A. Kirschel; pill millipede, M. Hedin; rhinoceros beetle, S. Yanoviak; butterfly, S. McCann)

Animal Examples Mammals have evolved many ways of protecting themselves, both from members of their own species and from potential predators. Wildebeest live in large herds and use speed as a primary means of defense. They also possess horns, which can be used as a last line of defense from predators, as well as for fighting members of their own species for access to mates.

Other animals use coverings as protection. Porcupines are covered in quills, which are actually keratinized hairs and deter predators. Armadillos are protected by a series of bony plates covering the back, head and tail. When attacked, armadillos roll into a tight ball, protecting the vulnerable belly. Lions may not seem as though they need protection, but the shaggy mane covering the neck of the male may act as protection when they engage in to-the-death fights over access to females.

Mammals use behaviors to protect themselves as well as modifications to their skin and protrusions from their skin. (Photographs courtesy of: wildebeest, Wikimedia, C. Rosenthal; porcupine, L. Parenteau; armadillo, A. Patterson, morgueFile.com; lion, B. Larison)

Tree pangolin
Manis tricuspis

Phylum:	Chordata	**Family:**	Manidae
Class:	Mammalia	**Genus:**	*Manis*
Order:	Pholidota	**Species:**	*M. tricuspis*

Photograph courtesy of E. Simpson

Habitat & Climate There are eight species of living pangolins, all within the genus *Manis*. The tree pangolin, a small African species, inhabits areas with suitable tree cover such as woody savannahs and the lowland tropical rainforests of western Africa. The location for the project is the Kakum National Park in Ghana, located in the heart of West Africa and home to lush tropical rainforest in less than 350 km² (144.8 mi²), containing many of the world's most endangered plant and animal species. The average yearly air temperature ranges from 26 to 33°C (~79–91°F); the park receives a high amount of precipitation, an average yearly amount of 1100 mm/yr (43.3 in.), corresponding to ~57 days of rain. The park's morning humidity averages 96% while the evening average is 79%. The average hourly direct normal radiation range is 425 to 75 Wh/m², and the average hourly direct normal illumination range is 35,000 to 5000 lux.

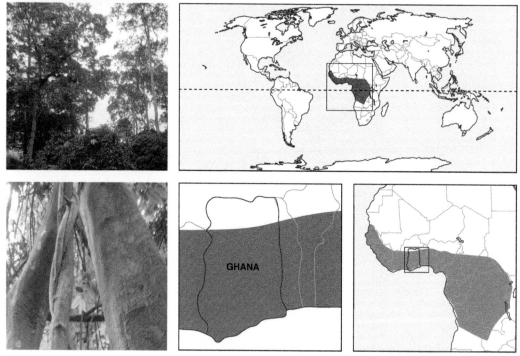

Habitat images of Kakum National Park, Ghana, West Africa. Tree pangolin distribution map. (Photographs courtesy of I. Mazzoleni)

Animal Physiological, Behavioral & Anatomical Element
Tree pangolins weigh 4.5–14 kg (10–31 lbs) and have body lengths
of 31–45 cm (12–18 in.). The pangolin's sides and back are covered with large, overlapping scales composed of keratin, similar
to human nails. When threatened, pangolins curl into a tight ball,
so the scales provide a layer of defense from predators and sting-
ing insects, as they feed exclusively on ants and termites. Due to
their lack of teeth, they possess tubular snouts and long tongues.
At rest, the tongue is drawn into the throat by long powerful
muscles originating from the sternum. Unusually large mucus-
producing salivary glands lubricate the tongue and provide an
adhesive surface on which ants can be trapped. Tree pangolins
spend much of their time in the trees and have developed a large,
muscular tail that serves both as a counterbalance and a fifth limb
to aid in climbing and hanging from branches.

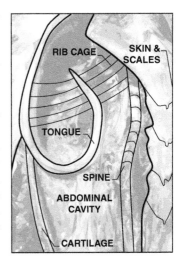

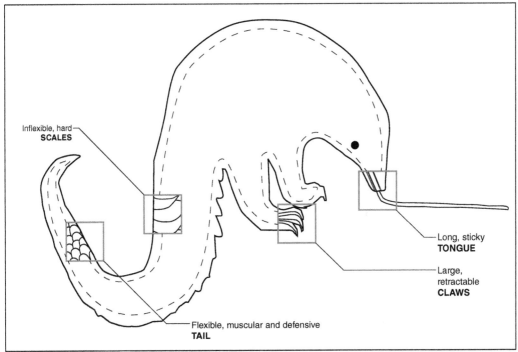

The main morphological characteristics of the tree pangolin include protective scales, large retractable claws for
climbing and opening termite nests, and a long, sticky tongue which allows the pangolin to reach deep into termite
and ant mounds. The scales are grooved or corrugated on their surface. In burrowing pangolins, this protects against
abrasive damage from dirt and rocks. In tree pangolins, it may function to protect from scratching branches and
thorns. (Photograph courtesy of K. Benirschke; project team: R. Ferrari & T. Carpentier)

Interface between the Skin & External World Like all mammals, the pangolin's skin is composed of three layers: the hypodermis, dermis, and epidermis. However, the overall physical appearance of the pangolin is dominated by large, hardened, plate-like scales that protect the animal from predators and prey. The scales are actually keratinized cells that grow from raised papillae protruding from the surface of the skin. The scales of newborn pangolins are soft but harden as they mature. The exposed outer and inner surfaces of the scales may fray through wear, but are replaced by newly keratinized cells from the middle layer.

Critically, the pangolin's skin system provides a hard protective system that maintains sufficient flexibility for the animal to climb or dig. The scales have a corrugated surface, which prevents excessive localized damage from everyday wear and tear.

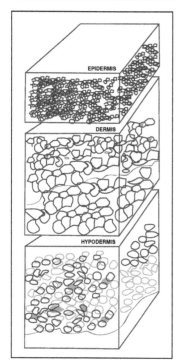

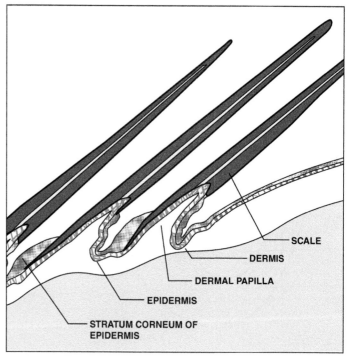

The illustration shows a detail of a cluster of the pangolin's scales. The skin and scale section show the connection of the scale to the skin and the raised papillae from which the scale growth occurs. (Photograph courtesy of K. Benirschke; diagram adapted from D. Visset; copyright credit Editiones Belin, 2006)

Interrelationship between the Skin & Internal Systems   The pangolin is often compared to a walking pine cone or globe artichoke. Although they tend to move at a slow pace, the flexibility afforded by scale armor, as opposed to a solid shell, means that the pangolin can move much faster when needed. When threatened, they are able to rear up on their hind limbs while utilizing the tail for balance. This increases the size of the pangolin and exposes the large claws in a warning threat display.

When threatened and unable to defend itself, the pangolin will roll up into a ball, using its prehensile tail to cover its face. Pangolins can also produce a noxious secretion from anal glands, which may also scare off a predator. In conjunction with protective eyelids, the pangolin is able to resist the attacks of soldier ants and termites and feed with relative ease.

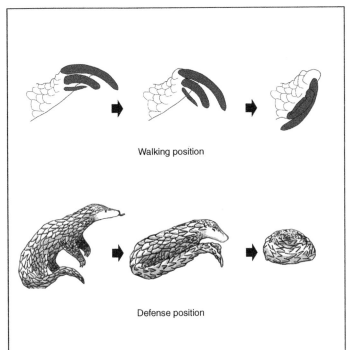

Walking position

Defense position

The diagram illustrates the pangolin's retractable claws, which allow them to walk on their front knuckles with the claws tucked underneath to protect them from wear, similar to giant anteaters. The claws also provide them with an additional protective element. The pangolin's tail plays a crucial role in movement and protection. (Photographs courtesy of: climbing, A. Kirschel; hanging, D. Ellis)

Proto-Architectural Project This project explores the duality of the protective and flexible qualities of the pangolin skin system, juxtaposing a flexible skin made up of rigid scales. Using Kakum National Park in Ghana as an initial project location, the building program evolves into an emergency pavilion for post-natural disaster or war zones. The project focuses on a premise to construct and develop a deployable and temporary structure that would adapt to varying site conditions and terrains. In the given scenario the primary interest is to establish temporary shelters to assist medical teams. The panels are held together using the principle of tensegrity, to provide the structural capacity necessary to support the panels and produce an inhabitable environment. In this instance the panels are lifted and maintained in their positions through tension and compression applied through the structural posts and cables.

An initial study model explored the flexible geometry of origami paper folding, especially with regard to the notion of expandable geometry. The examination led to the development of a final study model of a shelter which accommodates people and different programmatic activities. The final shelter design may be expanded and contracted on site to respond to different conditions such as light porosity, area, volume, and whatever degree of environmental protection is necessary.

Project Documentation   The shelter is designed for transportability in a series of basic pieces: photovoltaic panels, flexible translucent fabric, and a connecting umbrella-like structure. Whereas the pangolin's skin system provides protection against prey and predators, the adaptability of this shelter focuses on providing protection from undesirable environmental conditions. To address the necessity of energy, the design clads the modules with photovoltaic panels so that energy becomes readily available.

Assembly of the shelter is a multi-step process that begins by expanding the structural umbrella to stretch the protective fabric. Next, the photovoltaic panels are attached to the basic unit modules, which are then attached to form a stable arch, creating structural rigidity for the shelter. Finally, rows of unit modules are repeated until the desired volumetric space is achieved.

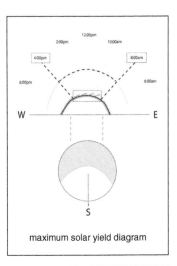

maximum solar yield diagram

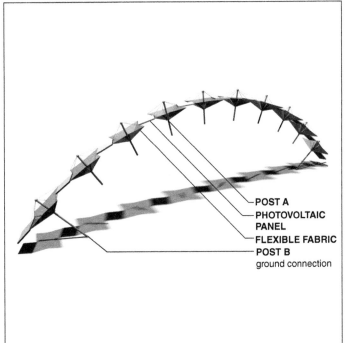

POST A
PHOTOVOLTAIC PANEL
FLEXIBLE FABRIC
POST B
ground connection

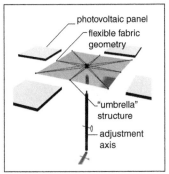

photovoltaic panel
flexible fabric geometry
"umbrella" structure
adjustment axis

A series of assembled modules lock in place to form a stable arch. The rigid photovoltaic panels are oriented to act as a protective shading element as well as to capture solar energy for use in the shelter. Right middle: singular assembled module. Right bottom: exploded diagram of an assembled module.

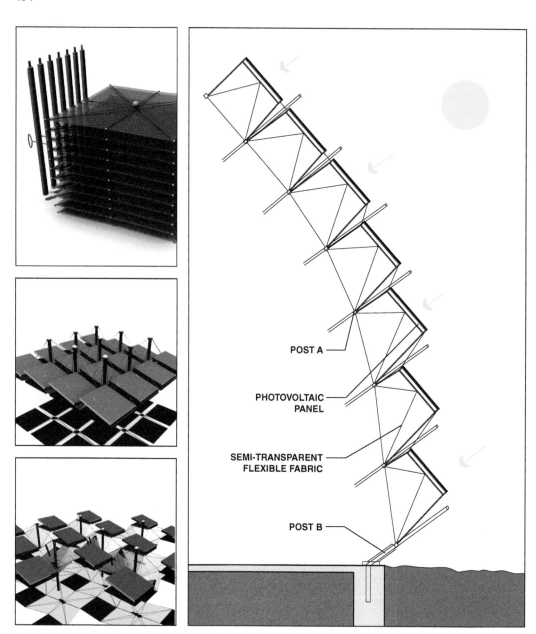

POST A

PHOTOVOLTAIC
PANEL

SEMI-TRANSPARENT
FLEXIBLE FABRIC

POST B

The diagrams illustrate the flexibility derived through the shelter's kit-of-parts assembly. En route to a disaster, the shelter elements are disassembled and shipped in small containers to make transport more effective. By opening or closing the assembled modules, the shelter envelope can be expanded or contracted to adapt to surface area and volume needs.

Inspired by the protective and flexible qualities of the pangolin skin system, the nomadic modular shelter provides protection from environmental elements while producing electricity for its inhabitants. The translucent envelope made of fabric allows an abundance of light and natural ventilation to pierce through, while maintaining a rigid tactility through principles of tensegrity.

Animal Examples Mammals, including humans, keep cool in a number of ways. Our skin is hairless, allowing air to flow freely over it. Sweating causes body heat to be lost via evaporative cooling. Tanning results from increased melatonin production to protect from UV radiation. Warthogs are also mostly hairless. They spend the hottest parts of the day in cool underground shelters and may wallow in mud to provide additional cooling. Spotted hyenas are covered in fur. Their fur is pale in color to reflect UV radiation and minimize overheating.

When hyenas do overheat, they seek shade and pant, like domestic dogs, to lose heat via evaporative cooling from the tongue. Rhinoceroses possess thick, dark, hairless skin. Water loss and UV damage are minimized by the densely keratinized epidermis. Rhinos, too, may wallow to further cool themselves.

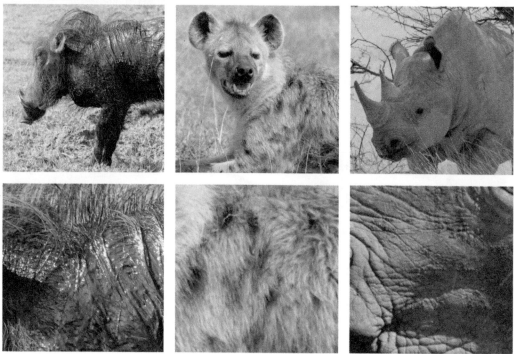

Animals have to protect themselves from environmental conditions, such as heat. They do this through sweating, panting, and spending time in the shade and other cool places. (Photographs courtesy of: human skin, S. Yelisee; warthog, spotted hyenas and rhinoceros, B. Larison)

Hippopotamus
Hippopotamus amphibius

Phylum:	Chordata	Family:	Hippopotamidae
Class:	Mammalia	Genus:	*Hippopotamus*
Order:	Artiodactyla	Species:	*H. amphibius*

Photograph courtesy of B. Larison

198

Habitat & Climate   There are two species in the Hippopotamidae family. The largest species, the common hippo, occurs only in grassland areas of subsaharan Africa where still water can be found. Hippos occupy an open stretch of water where they can rest during the day. At night, they leave the water to graze on grasses.

The Okavango Delta of Botswana is an optimal common hippo habitat. The Delta is an oasis in an otherwise arid landscape with a predominantly subtropical climate. Average rainfall is 450 mm (17.7 in.), most of which falls between December and March during heavy thunderstorms. Seasons vary from hot, wet summers during December through February to cold, dry winters in June through August. In the summer, temperatures reach as high as 40°C (104°F) with humidity levels between 50 and 80%. Night temperatures during winter may reach barely above freezing.

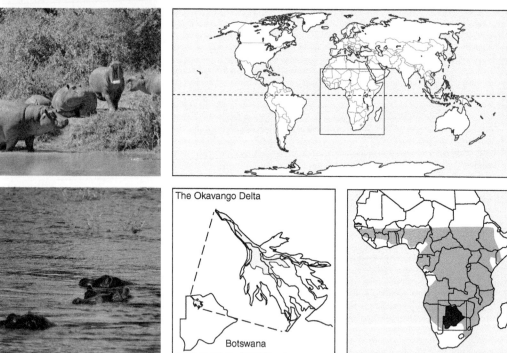

Botswana is dominated by the Kalahari Desert, covering 70% of its surface. The Okavango Delta in the Northwest offers a reprieve from the otherwise parched environment. During the Delta's annual flood, the water-covered area increases almost three-fold, from about 5000 km² (1930 mi²) to 6000–12,000 km² (2316 mi²–4634 mi²) during the winter months. (Photographs courtesy of B. Larison)

Animal Physiological, Behavioral & Anatomical Elements
The common hippo is characterized by a barrel-shaped torso, enormous mouth, large teeth, and relatively short legs ending in feet with webbed toes that make it easier to move underwater. The hippo's skin is almost hairless, with just a few bristles around the mouth and the tip of the tail, and is especially thick over the back and rump. Skin color varies from grayish-brown on the back to pink on the belly and around the eyes and the ears where the skin is thinnest. The hippo's thick skin is further protected by secreting a thick fluid from glands in the skin that protects against sunburn and keeps the coarse skin moist when the hippo is out of the water. Hippos also stay under water as a means of protecting the skin from sunburn and to keep cool. During hot days hippos come out of the water to graze only at night.

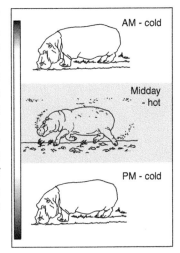

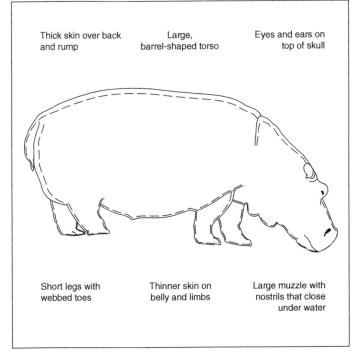

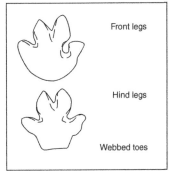

The eyes, ears and nostrils are placed high on the skull so that they do not submerge. The hippo's large muzzle has nostrils that close when underwater, allowing the hippo to fully submerge for several minutes at a time. The large body mass of the hippo allows it to walk underwater. It cannot float or swim. They move by bouncing off the riverbed floor and walking on the bottom. Webbing between the toes helps to propel them through the water. (Project team: S. Månsson & W. Raksaphon)

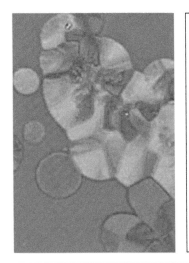

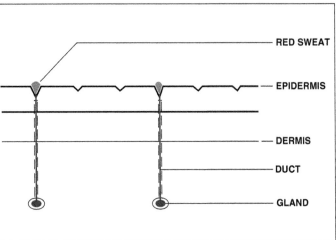

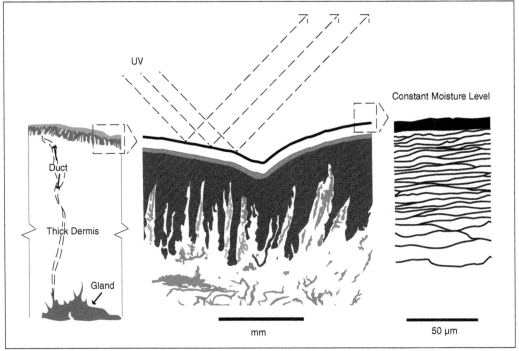

The thick skin consists of a dense sheet of collagen with fibers arranged in a matted, but regular, pattern giving great strength to the skin. Hippos have neither sweat nor sebaceous glands, but they have glands that produce a viscous red fluid, leading to the myth that hippos "sweat blood." Hippos secrete a natural sunscreen from glands deep within the dermis. This substance protects the thin epidermis from harmful UV rays, while also providing some antibacterial protection. (Micrograph courtesy of Reed et al. [2009]; diagram adapted from Luck and Wright [1963])

Interrelationship between the Skin & Internal Systems   The principal physiological functions of the hippo's skin are to control body temperature and regulate water loss. The hippo's semi-aquatic lifestyle allows it to control both simultaneously. By remaining submerged in water during the hottest parts of the day, hippos avoid body warming due to environmental factors. Hippos also have relatively dense bone, particularly in their limbs, to neutralize the natural buoyancy of the body and reduce the amount of effort and energy required to remain submerged. During the dry season when water levels are low, temperature regulation can occur via the activation of thermal windows. Thermal windows are areas of the body with momentarily higher temperatures than the surrounding body surface and ambient temperature. These windows of warmed skin lose heat to the environment, cooling the body in the process.

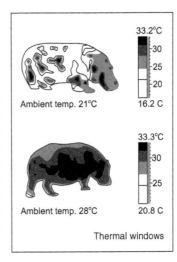

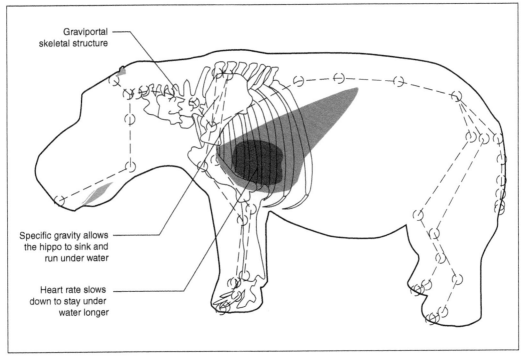

Hippos submerge with their ears pressed down flat and nostrils clamped shut. While underwater, their heart rate slows so they can stay underwater longer. As hippos feed on land, temperature regulation can be aided by flushing body heat to thermal windows in the skin. Thermal windows are increased in higher ambient temperatures. (Diagram adapted from Schneider and Kolter [2009])

Proto-Architectural Project   The project entitled "The Porous Skin" controls the intake of daylight while managing temperature during the different seasons. Inspiration for the design is found from the hippo's semi-aquatic lifestyle and the protective quality observed in its skin. The project's building envelope creates a comfortable indoor climate in both the warm summer and the colder winter months. The sliding panels regulate both the amount of daylight coming into the building and the water retained in the envelope assembly, which acts as a cooling element during the summer and a thermal mass in the winter. The permeable enclosure varies from a thin almost perforated envelope to a thicker dense wall, allowing for the transfer of heat from one layer of the façade to another. The various orientations of the opening within the wall unit control the direction of the light seeping into the building and enable control over the heat intake.

Hippo skin produces protective fluid in response to changing environmental conditions. The panelized façade system is created to accommodate the changing climate of the Okavango Delta which is characterized by warm and wet summer months and a cold and dry winter season. This design makes it possible for the building envelope to change through different conditions, always taking full advantage of the varying climate of the building site.

Project Documentation The building envelope is composed of movable panels shifting from one position to another. This creates a skin system that transitions from an entirely closed envelope that lets in very little daylight to an open envelope that allows the sun to heat up the interior of the building. This design accommodates the changing climate of the Okavango Delta.

When the panels are set to position A, water flows into the outer layer of the skin to cool the interior of the building during the hot and wet summer season. Opening the structure in the winter season with panels set to position B allows for more sunlight to permeate the façade, and helps warm the living spaces. Waiting until the peak of winter to set panels to an open position helps retain water between the layers of the envelope, functioning as a thermal mass which slowly releases stored heat within the building.

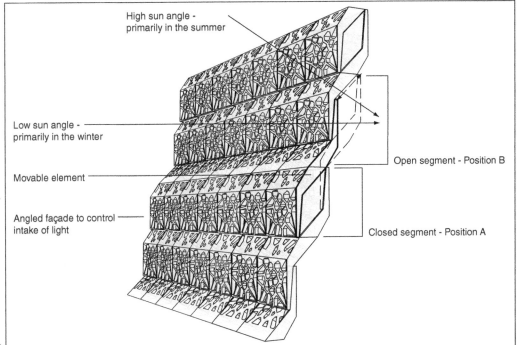

High sun angle - primarily in the summer

Low sun angle - primarily in the winter

Movable element

Angled façade to control intake of light

Open segment - Position B

Closed segment - Position A

The intent of the building envelope is to create a comfortable indoor climate in both the warm summer and the colder winter months. Sensors manage the sliding of the wall units controlling the overall amount of light directly hitting the floor plate, therefore regulating the temperature in specific areas of the building.

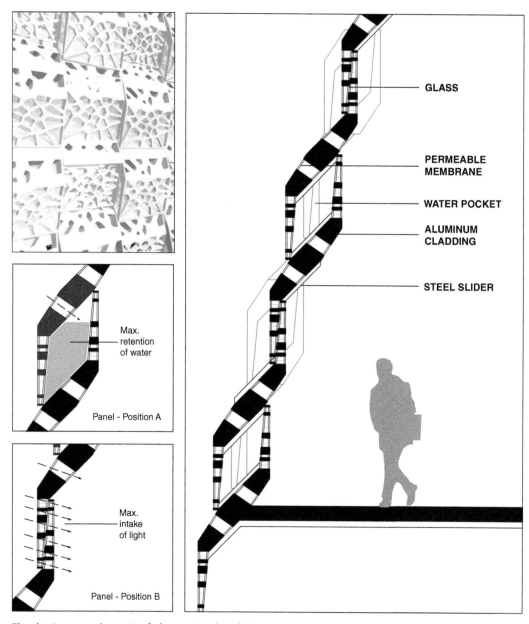

GLASS

PERMEABLE
MEMBRANE

WATER POCKET

ALUMINUM
CLADDING

STEEL SLIDER

Max.
retention
of water

Panel - Position A

Max.
intake
of light

Panel - Position B

The aluminum panels consist of a base unit with multiple openings. One half of the panels contain vertical apertures which let in large amounts of sunlight. The other half of the panels has fewer, slightly tilted openings, angled to capture rainwater. When panels slide into the closed position the angled apertures open and become permeable to both light and water, while the vertical openings remain as apertures, permeable only to light.

The panelized façade system creates a dense and layered, functional and ornamental building envelope. Filtering light through a permeable membrane drastically alters the ambience of the internal spaces as light glimmers among the internal surfaces fluctuating internal temperatures to comfortable levels. The porous and formed expression of the façade promotes a unique condition of shifting panels to establish an operable building that opens and closes as it registers the changes in the surrounding environment.

Endnotes

PART I

1. Biomimicry is a term coined by Janine M. Benyus in her book *Biomimicry. Innovation Inspired by Nature* (1997).

2. *Oxford Dictionary*: Prefixed to nouns and adjectives with the sense "earliest, original; at an early stage of development, primitive; incipient, potential."

3. Philip Drew *Touch This Earth Lightly: Glenn Murcutt in His Own Words"* (1999).

4. These data do not include species that are extinct in the wild. For further data on extinction rates see http://iucn.org/what/tpas/biodiversity/about/biodiversity_crisis/.

5. Ibid.

6. SCI-Arc Lecture, March 8, 2012.

7. Ibid.

8. Tom Wiscombe, "Beyond assemblies: System convergence and multi-materiality," *Bioinspir. Biomim.* 7 (2012), IOP Publishing.

9. "Materiomics is an emerging field of science that provides a basis for multiscale material system characterization, inspired

in part by natural, for example, protein-based materials." From *Materiomics: Biological Protein Materials, from Nano to Macro* S. Cranford and M.J. Buehler, Dovepress (2010).

PART II

[10] In humans, this organ serves as a protective barrier against microorganisms, physical damage, UV radiation, water loss, and the elements. It helps control body temperature through evaporative cooling, heat retention, radiation, convection, and conduction. It aids in fighting infections by providing an acidic environment and anti-microbial secretions and through initiating the body's immune response and the healing process.

[11] Many animal species are sexually dimorphic, meaning that males and females take different forms. Often, drab-colored females will be attracted to males that are vividly colored, because bright colors are thought to be an indicator of male health. An evolutionary advantage may be gained by females who mate with healthier males, as their offspring are likely to be healthier.

[12] In vertebrates, pigments are found in specialized branching cells called chromatophores. The distribution of the pigment within the cell is generally what causes dark coloration (when pigments are distributed throughout the cell) or light coloration (when they are concentrated in one portion of the cell). Coloration also has to do with the amount of pigment and the rate it is produced, or a change in the shape of the chromatophores. Invertebrates share the same pigments as vertebrates, but they are not always contained within specialized cells. Brown or black colors result from the presence of melanin; its stable chemical structure helps protect the skin from abrasion, and it absorbs UV light, which can damage tissue. Carotenoid pigments on their own provide yellow, orange, and red coloration; however, they are sometimes associated with different proteins, which can cause almost any kind of coloration. Carotenoids cannot be manufactured by animals, so they need to be obtained from plants. Finally, pterins are colorless, red, or yellow pigments that occur mostly in insects. When they do occur in vertebrates, pterins are often associated with melanin or carotenoids and contribute to yellow, orange, and red coloration.

[13] Some contain chemicals, such as guanine, that contribute to white coloration. Other chemicals reflect light and can work with chromatophores to produce iridescent colors. Animals that can rapidly change their coloration, such as fish, amphibians, and reptiles, are usually able to do so by regulating the concentration of pigments within chromatophores.

[14] Layers of air pockets or of certain kinds of particles in the skin occur in layers, and the interference between light waves is reflected in different layers. Interference occurs when light waves reflect off surfaces and coincide with each other, leading to a perception of brilliant color. The thickness and spacing of the layers determine which color will be seen. Iridescent colors are intense but change with the angle of view and can disappear entirely when viewed from a different direction. The most common iridescent colors are blue and green, but they can also be violet, gold, orange, and red. Often, pigment and light-scattering layers are combined to produce exceptionally pure colors. White is usually a structural color where all wavelengths of light are reflected from the surface.

[15] Additionally, birds use an oil secreted from the uropygial gland to preen feathers, a process that aids feather interlocking.

[16] A high salt concentration in the environment can cause water to be lost and desiccation to occur. Freshwater has a lower salt level than body tissues, causing water to seep into cells, ultimately causing them to burst. Aquatic organisms and even amphibians have various physiological mechanisms to maintain this balance.

Bibliography

Addington, M., and D. L. Schodek. 2004. *Smart Materials and Technologies in Architecture*. Oxford, UK: Architectural Press.

Alibardi, L. 2003. "Adaptation to the Land: The Skin of Reptiles in Comparison to That of Amphibians and Endotherm Amniotes." *Journal of Experimental Zoology Part B: Molecular and Developmental Evolution* 298B (1) (August 15): 12–41. doi:10.1002/jez.b.24.

Allen, G. R, and R. C. Steene. 1997. *Tropical Marine Life*. Singapore: Periplus Editions.

Allen, R., Ed. 2010. *Bulletproof Feathers: How Science Uses Nature's Secrets to Design Cutting-Edge Technology*. Chicago, IL: University of Chicago Press.

Amstrup, S. C., and D. P. DeMaster. 2003. "Polar Bear, *Ursus Maritimus*." *Wild Mammals of North America: Biology, Management, and Conservation* 587–610.

Angilletta, M. J., T. Hill, and M. A. Robson. 2002. "Is Physiological Performance Optimized by Thermoregulatory Behavior?: A Case Study of the Eastern Fence Lizard, *Sceloporus Undulatus*." *Journal of Thermal Biology* 27 (3): 199–204.

Anon. "ARKive—Violet-Tailed Sylph Videos, Photos and Facts—*Aglaiocercus Coelestis*." http://www.arkive.org.

Aranda, B., and C. Lasch. 2005. *Pamphlet Architecture 27: Tooling*. New York: Princeton Architectural Press.

Arnett, R. H., Jr., M. C. Thomas, P. E. Skelley, and J. H. Frank, Eds. 2002. *American Beetles*, Volume II: Polyphaga: Scarabaeoidea through Curculionoidea. Boca Raton, FL: CRC Press.

Ausubel, K., and J. P. Harpignies. 2004. *Nature's Operating Instructions: The True Biotechnologies*, San Francisco, CA: Sierra Club Books.

Avery, M., E. Tillman, and K. Krysko. 2009. "*Gopherus Polyphemus* (Gopher Tortoise), *Ctenosaura Similis* (Gray's Spiny-Tailed Iguana) Predation." USDA National Wildlife Research Center Staff Publications (January 1).

Baeyer, H. C. von. 1999. *Warmth Disperses and Time Passes: The History of Heat*. New York: Modern Library.

Baird, J. M. 1986. "A Field Study of Thermoregulation in the Carpenter Bee *Xylocopa Virginica Virginica* (Hymenoptera: Anthophoridae)." *Physiological Zoology* 157–168.

Banham, R. 1984. *Architecture of the Well-Tempered Environment*. 2nd ed. Chicago, IL: University of Chicago Press.

Banister, K. E., and A. Campbell, Eds. 1985. *The Encyclopedia of Aquatic Life*, illustrated ed. New York: Checkmark Books.

Banta, M. R., and D. W. Holcombe. 2002. "The Effects of Thyroxine on Metabolism and Water Balance in a Desert-Dwelling Rodent, Merriam's Kangaroo Rat (*Dipodomys Merriami*)." *Journal of Comparative Physiology B: Biochemical, Systemic, and Environmental Physiology* 172 (1): 17–25.

Bar-Cohen, Y. 2011. *Biomimetics: Nature-Based Innovation*. Boca Raton, FL: CRC Press.

Barker, G.M. 1999. *Naturalised Terrestrial Stylommatophora (Mollusca: Gastropoda). Fauna of New Zealand*. Canterbury, New Zealand: Manaaki Whenua Press.

Barr, R. A. 1927. "Memoirs: Some Notes on the Mucous and Skin Glands of *Arion Ater*." *Quarterly Journal of Microscopical Science* s2-71 (283) (December 1): 503–525.

Baskin, Y. 1998. *The Work of Nature: How the Diversity of Life Sustains Us*. Washington, DC: Island Press.

Beesley, P., Ed. 2010. *Kinetic Architectures and Geotextile Installations*. Cambridge, Ontario, Canada: Riverside Architectural Press.

Benirschke, K. 2007. White-Bellied Tree Pangolin. Comparative Placentation.

Benyus, J. M. 2002. *Biomimicry: Innovation Inspired by Nature*. New York: William Morrow Paperbacks.

Beukers, A. 2005. *Lightness*. 4th ed. Rotterdam, the Netherlands: 010 publishers.

Binetti, V. R., J. D. Schiffman, O. D. Leaffer, J. E. Spanier, and C. L. Schauer. 2009. "The Natural Transparency and Piezoelectric Response of the Greta Oto Butterfly Wing." *Integrative Biology* 1 (4): 324–329.

Blumberg, M. S. 2004. *Body Heat: Temperature and Life on Earth*. Cambridge, MA: Harvard University Press.

Born, M., F. Bongers, E. H. Poelman, and F. J. Sterck. 2010. "Dry-Season Retreat and Dietary Shift of the Dart-poison Frog *Dendrobates Tinctorius* (Anura: Dendrobatidae)." *Phyllomedusa* 9 (1):31–52.

Bracke, M.B.M. 2011. "Review of Wallowing in Pigs: Description of the Behaviour and Its Motivational Basis." *Applied Animal Behaviour Science* 132 (1-2) (June): 1–13.

Braddock, S. E., and M. O'Mahony. 2002. *SportsTech: Revolutionary Fabrics, Fashion and Design*. New York: Thames & Hudson.

Bradley, T. J. 2009. *Animal Osmoregulation*. New York: Oxford University Press.

Braungart, M. 2002. *Cradle to Cradle: Remaking the Way We Make Things*. New York: North Point Press.

Brebbia, C. A. 2006. *Design and Nature III: Comparing Design in Nature with Science and Engineering*. C. A. Brebbia, Ed. Southampton, UK: WIT Press.

———. 2008. *Design & Nature IV: Comparing Design in Nature with Science and Engineering*. Southampton, UK: WIT Press.

Brebbia, C. A., and M. W. Collins. 2004. *Design and Nature II: Comparing Design in Nature with Science and Engineering (Design and Nature)* illustrated ed. M. W. Collins, Ed. Southampton, UK: WIT Press /Computational Mechanics.

Breitkreutz, D., N. Mirancea, and R. Nischt. 2009. "Basement Membranes in Skin: Unique Matrix Structures with Diverse Functions?" *Histochemistry and Cell Biology* 132 (1) (July): 1–10.

Brodsky, L. M., and P. J. Weatherhead. 1984. "Behavioural Thermoregulation in Wintering Black Ducks: Roosting and Resting." *Canadian Journal of Zoology* 62 (7): 1223–1226.

Brown, G. Z., and M. DeKay. 2000. *Sun, Wind & Light: Architectural Design Strategies*, 2nd ed. New York: Wiley.

Brownell, B., Ed. 2010. *Transmaterial 3: A Catalog of Materials That Redefine Our Physical Environment*. New York: Princeton Architectural Press.

———. 2011. *Material Strategies: Innovative Applications in Architecture*. New York: Princeton Architectural Press.

Brusca, R. C., G. J. Brusca et al. 2002. *Invertebrates*, 2nd ed. Sunderland, MA: Sinauer Associates Incorporated.

Buehler, M. J., and S. Cranford. 2010. "Materiomics: Biological Protein Materials, from Nano to Macro." *Nanotechnology, Science and Applications* (November): 127.

Butcher, F. G. 1953. "Unusual Abundance of the Tree-Hopper *Umbonia Crassicornis* A & S." *Florida Entomologist* 36 (2) (May 31): 57–59.

Camazine, S., J.-L. Deneubourg, N. R. Franks, J. Sneyd, G. Theraula, and E. Bonabeau. 2003. *Self-Organization in Biological Systems*. New York: Princeton University Press.

Capra, F. 2008. *The Science of Leonardo: Inside the Mind of the Great Genius of the Renaissance*. Reprint. New York: Anchor.

Casey, T. M., and K. K. Casey. 1979. "Thermoregulation of Arctic Weasels." *Physiological Zoology*: 153–164.

Catchpoole, D. "Parrot Fashion." http://creation.mobi/parrot-fashion.

Chappell, M. A. 1982. "Temperature Regulation of Carpenter Bees (*Xylocopa Californica*) Foraging in the Colorado Desert of Southern California." *Physiological Zoology* 267–280.

Cheng, M. W., and R. L. Caldwell. 2000. "Sex Identification and Mating in the Blue-Ringed Octopus, *Hapalochlaena Lunulata*." *Animal Behaviour* 60 (1) (July): 27–33.

Chiesa, G. 2010. Biomimetica, tecnologia e innovazione per l'architettura. CELID.

Cisneros, A. B., and B. L. Goins, Eds. 2009. *Body Temperature Regulation*. Hauppage, NY: Nova Science Pub Inc.

Ciszek, D. 1999. *"Manis gigantea"* (On-line), Animal Diversity Web. http://animaldiversity.ummz.umich.edu/site/accounts/information/ Manis_gigantea.html

Clark, K. B., and M. Busacca. 1978. "Feeding Specificity and Chloroplast Retention in Four Tropical Ascoglossa, with a Discussion of the Extent of Chloroplast Symbiosis and the Evolution of the Order." *Journal of Molluscan Studies* 44 (3): 272–282.

Clarke, B. T. 1997. "The Natural History of Amphibian Skin Secretions, Their Normal Functioning and Potential Medical Applications." *Biological Reviews* 72 (3): 365–379.

Cleland, E. E. 2011. "Biodiversity and Ecosystem Stability." *Nature Education Knowledge* 2 (1): 2.

Clement, G., P. Rahm, and G. Borasi. 2007. *Environ(ne)ment: Approaches for Tomorrow*. Bilingual. Milan, Italy: Skira.

Cloudsley-Thompson, J. 2001. "Thermal and Water Relations of Desert Beetles." *Naturwissenschaften* 88 (11): 447–460.

Coineau, Y., and B. Kresling. 1989. Le invenzioni della natura e la bionica. San Paolo Edizioni.

Cook, A. 1987. "Functional Aspects of the Mucus-Producing Glands of the Systellommatophoran Slug, *Veronicella Floridana*." *Journal of Zoology* 211 (2): 291–305.

Corbellini, G. 2010. *Bioreboot: The Architecture of R&Sie{n}*. New York: Princeton Architectural Press.

Curtis, N. E., S. E. Massey, J. A. Schwartz, T. K. Maugel, and S. K. Pierce. 2005. "The Intracellular, Functional Chloroplasts in Adult Sea Slugs (*Elysia Crispata*) Come from Several Algal Species, and Are Also Different from Those in Juvenile Slugs." *Microscopy and Microanalysis* 11 (S02): 1194–1195.

Curtis, N. E., S. E. Massey, and S. K. Pierce. 2006. "The Symbiotic Chloroplasts in the Sacoglossan *Elysia Clarki* Are from Several Algal Species." *Invertebrate Biology* 125 (4) (November 1): 336–345.

Daniels, K. 1999. *Advanced Building Systems: A Technical Guide for Architects and Engineers*. Basel, Switzerland: Birkhäuser Architecture.

———. 2000. *Low-Tech Light-Tech High-Tech*. Basel, Switzerland: Birkhäuser.

Danks, H. V. 2007. "The Elements of Seasonal Adaptations in Insects." *The Canadian Entomologist* 139 (1): 1–44.

Davis, J. R., and D. F. DeNardo. 2008. "Water Storage Compromises Walking Endurance in an Active Forager: Evidence of a Trade-Off between Osmoregulation and Locomotor Performance." *Journal of Comparative Physiology A: Neuroethology, Sensory, Neural, and Behavioral Physiology* 194 (8): 713–718.

———. 2010. "Seasonal Patterns of Body Condition, Hydration State, and Activity of Gila Monsters (*Heloderma Suspectum*) at a Sonoran Desert Site." *Journal of Herpetology* 44 (1): 83–93.

Davis, W. 2009. *The Wayfinders: Why Ancient Wisdom Matters in the Modern World*. Toronto, Canada: House of Anansi Press.

DeMaster, D. P., and I. Stirling. 1981. *"Ursus Maritimus." Mammalian Species* 145 (7). http://swfsc.noaa.gov/publications/CR/1981/8111. pdf.

Denny, M. W. 1989. "Invertebrate Mucous Secretions: Functional Alternatives to Vertebrate Paradigms." *Symposia of the Society for Experimental Biology* 43:337.

Denny, M. W., and J. M. Gosline. 1980. "The Physical Properties of the Pedal Mucus of the Terrestrial Slug, *Ariolimax Columbianus." The Journal of Experimental Biology* 88 (1): 375–394.

Deyrup-Olsen, I., and H. Jindrova. 1996. "Product Release by Mucous Granules of Land Slugs: *Ariolima Columbianus* as a Model Species." *Journal of Experimental Zoology* 276 (6): 387–393.

Deyrup-Olsen, I., D. L. Luchtel, and A. W. Martin. 1983. "Components of Mucus of Terrestrial Slugs (Gastropoda)." *American Journal of Physiology-Regulatory, Integrative and Comparative Physiology* 245 (3): R448–R452.

Diao, Y. Y., and X. Y. Liu. 2011. "Mysterious Coloring: Structural Origin of Color Mixing for Two Breeds of *Papilio* Butterflies." *Optics Express* 19 (10): 9232–9241.

Dill, P. B. and L. Irving. 1964. Polar Biology. In *Handbook of Physiology*, Chapter 5. Bethesda, MD: American Physiological Society.

Dollens, D. 2005. *Digital-Botanic Architecture: D-B-A*. Santa Fe, NM: Lumen Books.

Dollens, D., Ed. 2004. *Genetic Architectures/Arquitecturas Genéticas*. Bilingual. SITES.

Drew, P. 2000. *Touch This Earth Lightly: Glenn Murcutt in His Own Words*. Australia: Duffy & Snellgrove.

Duffey, K. 2006. "Species Spotlight: Violet-Tailed Sylph." *Las Gralarius Foundation Inc. Newsletter* 1 (1): 2.

Duncan, F. D., and J. R. B. Lighton. 1994. "Water Relations in Nocturnal and Diurnal Foragers of the Desert Honeypot Ant *Myrmecocystus*: Implications for Colony-Level Selection." *Journal of Experimental Zoology* 270 (4) (November 15): 350–359.

Eckert, R., and D. Randall. 1988. *Animal Physiology: Mechanisms and Adaptations*. 3rd ed. Gordonsville, VA: W. H. Freeman & Co.

Egri, Á., M. Blahó, G. Kriska, R. Farkas, M. Gyurkovszky, S. AAkesson, and G. Horváth. 2012. "Polarotactic Tabanids Find Striped Patterns with Brightness and/or Polarization Modulation Least Attractive: An Advantage of Zebra Stripes." *The Journal of Experimental Biology* 215 (5): 736–745.

Elkins, N. 1983. *Weather and Bird Behaviour*. London: T & AD Poyser Ltd.

Farrant, P. A. 1999. *Color in Nature: A Visual and Scientific Exploration*. New York: Sterling Publishing.

Faulkner, D. J., and M. T. Ghiselin. 1983. "Chemical Defense and Evolutionary Ecology of Dorid Nudibranchs and Some Other Opisthobranch Gastropods." *Marine Ecology Progress Series. Oldendorf* 13 (2): 295–301.

Ferino-Pagden, S. 2011. *Arcimboldo. Artista Milanese Tra Leonardo e Caravaggio*. Milan, Italy: Skira.

Finsterwalder, R., Ed. 2011. *Form Follows Nature: A History of Nature as Model for Design in Engineering, Architecture and Art*. New York: Springer Vienna Architecture.

Firth, B. T., J. S. Turner, and C. L. Ralph. 1989. "Thermoregulatory Behaviour in Two Species of Iguanid Lizards (*Crotaphytus Collaris* and *Sauromalus Obesus*): Diel Variation and the Effect of Pinealectomy." *Journal of Comparative Physiology* B 159 (1): 13–20.

Forbes, A. A., and B. A. Krimmel. 2010. "Evolution Is Change in the Inherited Traits of a Population through Successive Generations." *Nature Education Knowledge* 1 (10): 6.

Forbes, P. 2006. *The Gecko's Foot: Bio-Inspiration: Engineering New Materials from Nature*. New York: W. W. Norton & Company.

Forseth, I. 2010. "Terrestrial Biomes." *Nature Education Knowledge* 1 (8): 12.

Freeman, S. 2010. *Biological Science*. 4th ed. Reading, MA: Benjamin Cummings.

Frisch, K. Von. 1974. *Animal Architecture*. Boston, MA: Harcourt.

Furskin Identification. "*Uncia uncia*" Fursin.cz http://www.furskin.cz/overview.php?furskin=Uncia%20uncia

Gehler, J., M. Cantz, J. F. O'Brien, M. Tolksdorf, and J. Spranger. 1975. "Mannosidosis: Clinical and Biochemical Findings." *Birth Defects Original Article Series* 11 (6): 269–272.

Gerken, M. 2010. "Relationships between Integumental Characteristics and Thermoregulation in South American Camelids." *Animal* 4 (09): 1451–1459.

Gessaman, J. A. 1978. "Body Temperature and Heart Rate of the Snowy Owl." *The Condor* 80 (2): 243–245.

Gevorkian, P. 2006. *Sustainable Energy System Engineering: The Complete Green Building Design Resource*. New York: McGraw-Hill Professional.

Gibson, P. E. 1976. "Quantitative Analysis of the Major Subdeterminants of Hepatitis B Surface Antigen." *The Journal of Infectious Diseases* 134 (6) (December): 540–545.

Gill, J. 1993. "Influence of Environmental Factors on the Deep and Skin Temperature in the European Bison, *Bison Bonasus* (L.)." *Comparative Biochemistry and Physiology. Comparative Physiology* 106 (4) (December): 653–661.

Gillespie, L. J. 1997. "*Omphalea* (Euphorbiaceae) in Madagascar: A New Species and a New Combination." *Novon* 7: 127–136.

Gissen, D. 2009. *Subnature: Architecture's Other Environments*. New York: Princeton Architectural Press.

Gordon, D. G., and Western Society of Malacologists. 2002. *Field Guide to the Slug*. Seattle, WA: Sasquatch Books.

Gould, J. L., and C. Grant Gould. 2007. *Animal Architects: Building and the Evolution of Intelligence*. New York: Basic Books.

Gray, D. R. 1993. "Behavioural Adaptations to Arctic Winter: Shelter Seeking by Arctic Hare (*Lepus Arcticus*)." *Arctic*: 340–353.

Greene, Brian. 2000. *The Elegant Universe: Superstrings, Hidden Dimensions, and the Quest for the Ultimate Theory*. New York: Vintage Books.

Grondzik, W. T., A. G. Kwok, B. Stein, and J. S. Reynolds. 2009. *Mechanical and Electrical Equipment for Buildings*. 11th ed. New York: Wiley.

Gruber, P. 2010. *Biomimetics in Architecture: Architecture of Life and Buildings*. New York: Springer Vienna Architecture.

Gruber, P., D. Bruckner, C. Hellmich, H.-B. Schmiedmayer, H. Stachelberger, and I. C. Gebeshuber, Eds. 2011. *Biomimetics-Materials, Structures and Processes: Examples, Ideas and Case Studies*. Berlin: Springer.

Gruber, P., and G. Jeronimidis. 2012. "Has Biomimetics Arrived in Architecture?" *Bioinspiration & Biomimetics* 7 (1) (March 1): 010201.

Grynsztejn, M., Ed. 2007. *Take Your Time: Olafur Eliasson*. New York: Thames & Hudson.

Grzimek, B. 1972. *Grzimek's Animal Life Encyclopedia: Volume 3 Mollusks and Echinoderms*. New York: Van Nostrand Reinhold.

Gvoždík, L. 2011. "Plasticity of Preferred Body Temperatures as Means of Coping with Climate Change?" *Biology Letters* 8 (2): 262–265.

Haber, M., S. Cerfeda, M. Carbone, G. Calado, H. Gaspar, R. Neves, V. Maharajan, et al. 2010. "Coloration and Defense in the Nudibranch Gastropod *Hypselodoris Fontandraui*." *The Biological Bulletin* 218 (2): 181–188.

Haeckel, E. 1974. *Art Forms in Nature*. Revised. Mineola, NY: Dover Publications.

Hamilton, P. A., and W. G. Wellington. 1981. "The Effects of Food and Density of the Movement of *Arion Ater* and *Ariolimax Columbianus* (Pulmonata: Stylommatophora) between Habitats." *Researches on Population Ecology* 23 (2): 299–308.

Händeler, K., Y. P. Grzymbowski, P. J. Krug, and H. Wägele. 2009. "Functional Chloroplasts in Metazoan Cells — A Unique Evolutionary Strategy in Animal Life." *Frontiers in Zoology* 6 (1): 28.

Hardy, S., A. Martin, and M. Poletto. 2008. *Environmental Tectonics: Forming Climatic Change*. Ed. Steve Hardy. London: Architectural Association Publications.

Harley, C. D. G., M. S. Pankey, J. P. Wares, R. K. Grosberg, and M. J. Wonham. 2006. "Color Polymorphism and Genetic Structure in the Sea Star *Pisaster Ochraceus*." *The Biological Bulletin* 211 (3): 248–262.

Hashimoto, K., Y. Saikawa, and M. Nakata. 2007. "Studies on the Red Sweat of the *Hippopotamus Amphibius.*" *Pure and Applied Chemistry* 79 (4): 507–518.

Heath, M. E. 1992. "Manis Temminckii." *Mammalian Species* (415): 1–5.

Heinrich, B., and S. L. Buchmann. 1986. "Thermoregulatory Physiology of the Carpenter Bee, *Xylocopa Varipuncta.*" *Journal of Comparative Physiology B: Biochemical, Systemic, and Environmental Physiology* 156 (4): 557–562.

Heinrich, B., and H. Esch. 1994. "Thermoregulation in Bees." *American Scientist* 82 (2): 164–170.

Heinrich, B. et al. 1981. *Insect Thermoregulation.* New York: Wiley Interscience.

Heinrich, B. 1999. *The Thermal Warriors: Strategies of Insect Survival.* Cambridge, MA: Harvard University Press.

Hensel, M., A. Menges, and M. Weinstock. 2010. *Emergent Technologies and Design: Towards a Biological Paradigm for Architecture.* New York: Routledge.

Heschong, L. 1979. *Thermal Delight in Architecture.* Cambridge, MA: The MIT Press.

Hill's Pet Products. 2004. *Hill's Atlas of Veterinary Clinical Anatomy.* Lenexa, KS: Veterinary Medicine Publishing

Honacki, J. H., K. E. Kinman, and J. W. Koeppl. 1982. *Mammal Species of the World.* Cambridge University Press.

Houck, L. D., and L. C. Drickamer, Eds. 1996. *Foundations of Animal Behavior: Classic Papers with Commentaries.* Chicago, IL: University of Chicago Press.

Howell, A. B., and I. Gersh. 1935. "Conservation of Water by the Rodent Dipodomys." *Journal of Mammalogy* 16 (1): 1–9.

Hoyo, J. del, A. Elliott, and J. Sargatal, Eds. 1999. *Handbook of the Birds of the World, Vol. 5: Barn Owls to Hummingbirds.* Barcelona, Spain: Lynx Edicions.

Huxley, R., Ed. 2007. *The Great Naturalists.* New York: Thames & Hudson.

Ingraham, C. 2006. *Architecture, Animal, Human: The Asymmetrical Condition.* New York: Routledge.

Iwamoto, L. 2009. *Digital Fabrications: Architectural and Material Techniques.* 144 p. New York: Princeton Architectural Press.

Johansen, K., and C. Bech. 1983. "Heat Conservation during Cold Exposure in Birds (vasomotor and Respiratory Implications)." *Polar Research* 1 (3): 259–268.

Jones, C. A., Ed. 2006. *Sensorium: Embodied Experience, Technology, and Contemporary Art.* Cambridge, MA: The MIT Press.

Jones, T. R. 1861. *General Outline of the Organization of the Animal Kingdom, and Manual of Comparative Anatomy.* London: John Van Voorst.

Jørgensen, C. B. 1949. "Permeability of the Amphibian Skin." *Acta Physiologica Scandinavica* 18 (2–3): 171–180.

Jose, S., E. J. Jokela, and D. L. Miller, Eds. 2006. *The Longleaf Pine Ecosystem: Ecology, Silviculture, and Restoration.* Berlin: Springer.

Juhani, P. 1996. "Gli Animali Architetti." *Ottagono: Rivista Trimestrale Di Architettura, Arredamento e Industrial Design* (117): 65–80.

Kieran, S., and J. Timberlake. 2003. *Refabricating Architecture: How Manufacturing Methodologies Are Poised to Transform Building Construction*. New York: McGraw-Hill Professional.

Kimsey, L. S., and R. M. Bohart. 1991. *The Chrysidid Wasps of the World*. New York: Oxford University Press.

Kolarevic, B., Ed. 2006. *Architecture in the Digital Age*. New York: Taylor & Francis.

Kolarevic, B., and A. Malkawi, Eds. 2005. *Performative Architecture: Beyond Instrumentality*. New York: Routledge.

Krebs, C. J. 2001. *Ecology: The Experimental Analysis of Distribution and Abundance*. 5th ed. Reading, MA: Benjamin Cummings.

Kwinter, S. 2008. *Far from Equilibrium: Essays on Technology and Design Culture*. Cynthia Davidson, Ed. Barcelona, Spain: Actar.

Lally, S., and J. Young, Eds. 2007. *Softspace: From a Representation of Form to a Simulation of Space*. New York: Routledge.

Lally, S., Ed. 2009. *Energies:New Material Boundaries: Architectural Design*. New York: Wiley.

Larivière, S. 2002. "*Vulpes Zerda.*" *Mammalian Species*: 1–5.

Laurenza, Domenico. 2007. *Leonardo on Flight*. Baltimore, MD: The Johns Hopkins University Press.

Lavers, C. 2001. *Why Elephants Have Big Ears: Understanding Patterns of Life on Earth*. New York: St. Martin's Press.

Lecointre, G., and H. Le Guyader. 2001. *Classification phylogénétique du vivant*. Paris: Belin.

Lees, D. C, and N. G Smith. 1991. "Foodplant Associations of the Uraniinae (Uraniidae) and Their Systematic, Evolutionary, and Ecological Significance." *Journal of the Lepidopterists Society* 45 (4): 296–347.

Levine, J. M, and J. HilleRisLambers. 2009. "The Importance of Niches for the Maintenance of Species Diversity." *Nature* 461 (7261): 254–257.

Li, D., and L. D. Graham. 2007. "Epidermal Secretions of Terrestrial Flatworms and Slugs: *Lehmannia Valentiana* Mucus Contains Matrilin-like Proteins." *Comparative Biochemistry and Physiology Part B: Biochemistry and Molecular Biology* 148 (3): 231–244.

Lin, L. H., D. T. Edmonds, and F. Vollrath. 1995. "Structural Engineering of an Orb-Spider's Web." Published Online: 12 January 1995; (6510) (January 12): 146–148.

Liu, K., and L. Jiang. 2011. "Bio-Inspired Design of Multiscale Structures for Function Integration." *Nano Today* 6 (2) (April): 155–175.

Lovegrove, B. G. 2005. "Seasonal Thermoregulatory Responses in Mammals." *Journal of Comparative Physiology B: Biochemical, Systemic, and Environmental Physiology* 175 (4): 231–247.

Lowe, S. J. 2010. "Behavioral Responses of Moose (*Alces Alces*) to Ambient Temperature: Is There Evidence for Behavioral Thermoregulation?" Ontario, Canada: Trent University.

Luck, C. P., and P. G. Wright. 1964. "Aspects of the Anatomy and Physiology of the Skin of the Hippopotamus (*H. Amphibius*)." *Experimental Physiology* 49 (1) (January 1): 1–14.

Luehtel, D. L., I. Deyrup-Olsen, and A. W. Martin. 1991. "Ultrastructure and Lysis of Mucin-Containing Granules in Epidermal Secretions of the Terrestrial Slug *Ariolimax columbianus* (Mollusca: Gastropoda: Pulmonata)." *Cell and Tissues Research*, 266 (2): 375–383.

Lupton, E., and J. Tobias. 2002. *Skin: Surface, Substance + Design*. New York: Princeton Architectural Press.

MacKay, D. J. C. 2009. *Sustainable Energy—Without the Hot Air*. Chicago, IL: UIT Cambridge Ltd.

Madison, K. C. 2003. "Barrier Function of the Skin: 'La Raison D'être' of the Epidermis." *The Journal of Investigative Dermatology* 121 (2) (August): 231–241.

Magee, J. 2010. *Art and Nature: Three Centuries of Natural History Art from Around the World*. Vancouver, Canada: Greystone Books.

Maloiy, G. M. O., J. M. Z. Kamau, A. Shkolnik, M. Meir, and R. Arieli. 1982. "Thermoregulation and Metabolism in a Small Desert Carnivore: The Fennec Fox (*Fennecus zerda*)(*Mammalia*)." *Journal of Zoology* 198 (3): 279–291.

Martinez-Palomo, A., D. Erlij, and H. Bracho. 1971. "Localization of Permeability Barriers in the Frog Skin Epithelium." *The Journal of Cell Biology* 50 (2): 277–287.

Matério. 2005. *Material World 2: Innovative Materials for Architecture and Design*. Basel, Switzerland: Birkhäuser.

May, M. L. 1979. "Insect Thermoregulation." *Annual Review of Entomology* 24 (1): 313–349.

Mazzoleni, I., P. Ra, A. Barthakur, S. Price, V. Zajfen, S. Varma, B. Mehlomakulu, H. Portillo, S. Milner, and S. Proudian. 2008. "Ecosystematic Restoration: A Model Community at Salton Sea." In I:201–211. Southampton, UK: WIT Press.

Mazzoleni, Ilaria. 2010. "Involucri Biomimetici." Disegnare Con.

———. 2011a. "Biomimetic Envelopes: Investigating Nature to Design Buildings." In Proceedings of the First Annual Biomimicry in Higher Education Webinar.

———. 2011b. "A Zoological Approach to Architecture." *Domus*, February.

McGinnis, S. M. 1970. "Flexibility of Thermoregulatory Behavior in the Western Fence Lizard *Sceloporu Occidentalis*." *Herpetologica*: 70–76.

McGlynn, T. 2010. "Effects of Biogeography on Community Diversity." *Nature Education Knowledge* 1 (8): 32.

McQuaid, M. 2005. *Extreme Textiles: Designing for High Performance*. New York: Princeton Architectural Press.

Meadows, D H. 2008. *Thinking in Systems: A Primer*. White River Junction, VT: Chelsea Green Publishing.

Meredith, M. 2008. *From Control to Design: Parametric/ Algorithmic Architecture*. Ed. M. Meredith, Aranda-lasch, and M. Sasaki. Barcelona, Spain: Actar.

Mishra, C., P. Allen, T. O. M. Mccarthy, M. D. Madhusudan, Ag. Bayarjargal, and H. H. T Prins. 2003. "The Role of Incentive Programs in Conserving the Snow Leopard." *Conservation Biology* 17 (6) (December 1): 1512–1520.

Moe, K. 2010. *Thermally Active Surfaces in Architecture*. New York: Princeton Architectural Press.

Mora, C., D. P Tittensor, S. Adl, A. G. B. Simpson, and B. Worm. 2011. "How Many Species Are There on Earth and in the Ocean?" *PLoS Biology* 9 (8).

Morrison, P., M. Rosenmann, and J. A. Estes. 1974. "Metabolism and Thermoregulation in the Sea Otter." *Physiological Zoology* 47 (4): 218–229.

Moussavi, F. 2009. *The Function of Form*. Ed. D. Lopez, G. Ambrose, B. Fortunato, R. R. Ludwig, and A. Schricker. unknown. Barcelona, Spain: Actar.

Munari, B. 1981. *Da Cosa Nasce Cosa: Appunti Per Una Metodologia Progettuale*. Rome: Laterza.

Nagel, T. 1974. "What Is It Like to Be a Bat?" *The Philosophical Review* 83 (4): 435–450.

Nagy, K. A., and M. J. Gruchacz. 1994. "Seasonal Water and Energy Metabolism of the Desert-Dwelling Kangaroo Rat (*Dipodomys Merriami*)." *Physiological Zoology*: 1461–1478.

Naidu, S. G. 2001. "Water Balance and Osmoregulation in *Stenocara Gracilipes*, a Wax-Blooming Tenebrionid Beetle from the Namib Desert." *Journal of Insect Physiology* 47 (12) (December): 1429–1440.

Naidu, S. G. 2008. "Why Does the Namib Desert Tenebrionid *Onymacris unguicularis* (Coleoptera: Tenebrionidae) Fog-Bask?" *European Journal of Entomology* 105 (5): 829–838.

Nerdinger, W., Ed. 2001. *Frei Otto. Complete Works*. 2005. Basel, Switzerland: Birkhäuser Architecture.

Nicholson, K. E., L. J. Harmon, and J. B. Losos. 2007. "Evolution of Anolis Lizard Dewlap Diversity." *PLoS* One 2 (3): e274.

NOAA. "National Weather Service." Last modified July 27, 2011. http://www.weather.gov/

Noirard, C., M. Le Berre, R. Ramousse, and J. P. Lena. 2008. "Seasonal Variation of Thermoregulatory Behaviour in the Hippopotamus (*Hippopotamus Amphibius*)." *Journal of Ethology* 26 (1): 191–193.

Noonan, B. P, and P. Gaucher. 2006. "Refugial Isolation and Secondary Contact in the Dyeing Poison Frog *Dendrobates Tinctorius*." *Molecular Ecology* 15 (14): 4425–4435.

Nørgaard, T., and M. Dacke. 2010. "Fog-Basking Behaviour and Water Collection Efficiency in Namib Desert Darkling Beetles." *Frontiers in Zoology* 7 (1) (July 16): 23.

Nowak, R. M. 1999. *Walker's Mammals of the World*. 6th ed. Baltimore, MD: The Johns Hopkins University Press.

Onna, E. van. 2003. *Material World: Innovative Structures and Finishes for Interiors*. Basel, Switzerland: Birkhäuser.

Otto, F., and B. Rasch. 1996. *Finding Form: Towards an Architecture of the Minimal*. Fellbach, Germany: Axel Menges.

Pagès, E. 1975. "Eco-Ethological Study of *Manis Tricuspis* by Radio Tracking." *Mammalia* 39 (4): 613–641.

Pallasmaa, J. 2005. *The Eyes of the Skin: Architecture and the Senses*. 2nd ed. Chichester, UK: Wiley.

Parker, A. R., and C. R. Lawrence. 2001. "Water Capture by a Desert Beetle." *Nature* 414 (6859) (November 1): 33–34.

Patterson, M. 2011. *Structural Glass Facades and Enclosures*. New York: Wiley.

Pawlyn, M. 2011. *Biomimicry in Architecture*. New Castle upon Tyne, UK: RIBA Enterprises.

Pearce, P. 1980. *Structure in Nature Is a Strategy for Design*. Cambridge, MA: The MIT Press.

Peters, T. 2012. *Experimental Green Strategies: Redefining Ecological Design Research - Architectural Design*. 2nd ed. New York: Wiley.

Peterson, C. H. 1991. "Intertidal Zonation of Marine Invertebrates in Sand and Mud." *American Scientist* 79 (3): 236–249.

Piper, R. 2007. *Extraordinary Animals: An Encyclopedia of Curious and Unusual Animals*. Westport, CT: Greenwood Pub Group.

Pramatarova, L., Ed. 2011. *On Biomimetics*. Rijeka, Croatia: InTech.

Prestrud, P. 1991. "Adaptations by the Arctic Fox (*Alopex Lagopus*) to the Polar Winter." *Arctic*: 132–138.

Prestrud, P., and K. Nilssen. 1992. "Fat Deposition and Seasonal Variation in Body Composition of Arctic Foxes in Svalbard." *The Journal of Wildlife Management*: 221–233.

Proksch, E., J. M. Brandner, and J.-M. Jensen. 2008. "The Skin: An Indispensable Barrier." *Experimental Dermatology* 17 (12) (December): 1063–1072.

Prum, R. O., T. Quinn, and R. H. Torres. 2006. "Anatomically Diverse Butterfly Scales All Produce Structural Colours by Coherent Scattering." *Journal of Experimental Biology* 209 (4): 748–765.

Purvis, A., and A. Hector. 2000. "Getting the Measure of Biodiversity." *Nature* 405 (6783): 212–219.

Rahm, P. *Architecture Météorologique*. Paris: Archibooks.

Reed, E. J., L. Klumb, M. Koobatian, and C. Viney. 2009. "Biomimicry as a Route to New Materials: What Kinds of Lessons Are Useful?" *Philosophical Transactions of the Royal Society A: Mathematical, Physical and Engineering Sciences* 367 (1893): 1571–1585.

Reidenberg, J. S. 2007. "Anatomical Adaptations of Aquatic Mammals." *The Anatomical Record: Advances in Integrative Anatomy and Evolutionary Biology* 290 (6): 507–513.

Reis, R. L., and S. Weiner. 2004. *Learning from Nature How to Design New Implantable Biomaterials: From Biomineralization Fundamentals to Biomimetic Materials and Processing Routes.* Berlin: Springer.

Reiser, J. 2006. *Atlas of Novel Tectonics.* New York: Princeton Architectural Press.

Rudman, W.B., 2006. *Elysia crispata* (Morch, 1863). Sea Slug Forum. Australian Museum, Sydney. http://www.seaslugforum.net/factshee t/elyscris

Rundel, P. W., and A. C. Gibson. 1996. *Ecological Communities and Processes in a Mojave Desert Ecosystem.* Cambridge University Press.

Ryn, S. Van der, and S. Cowan. 2007. *Ecological Design*, Tenth Anniversary Edition. Washington, DC: Island Press.

Salvia, V. R., and M. Levi Giuseppe. 2009. *Il Progetto Della Natura. Gli Strumenti Della Biomimesi Per Il Design.* Rome: Franco Angeli.

Saikawa, Y., K. Hashimoto, M. Nakata, M. Yoshihara, K. Nagai, M. Ida, and T. Komiya. 2004. "Pigment Chemistry: The Red Sweat of the Hippopotamus." *Nature* 429 (6990): 363–363.

Saikawa, Y., K. Moriya, K. Hashimoto, and M. Nakata. 2006. "Synthesis of Hipposudoric and Norhipposudoric Acids: The Pigments Responsible for the Color Reaction of the Red Sweat of *Hippopotamus Amphibius.*" *Tetrahedron Letters* 47 (15): 2535–2538.

Schittich, C., Ed. 2002. *In Detail: Building Skins.* Basel, Switzerland: Birkhäuser Architecture.

Schmidt-Nielsen, K., E. C. Crawford, and H. T. Hammel. 1981. "Respiratory Water Loss in Camels." *Proceedings of the Royal Society of London. Series B. Biological Sciences* 211 (1184): 291–303.

Schmidt-Nielsen, K. 1972. *How Animals Work.* Cambridge University Press.

———. 1984. *Scaling: Why Is Animal Size So Important?* Cambridge University Press.

———. 1997. *Animal Physiology: Adaptation and Environment.* 5th ed. Cambridge University Press.

Schneider, M., L. Kolter, and C. Zoo. "Visualisation of Body Surfaces Specialized for Heat Loss by Infrared Thermography." http://www. thermotec-fi scher.de/pdf/Poster_Schneider_IZW_2009.pdf.

Scimone, M. Lucila, M. Srivastava, G. W. Bell, and P. W. Reddien. 2011. "A Regulatory Program for Excretory System Regeneration in Planarians." *Development* 138 (20) (October 15): 4387–4398.

Seago, A. E., P. Brady, J. P. Vigneron, and T. D. Schultz. 2009. "Gold Bugs and Beyond: A Review of Iridescence and Structural Colour Mechanisms in Beetles (Coleoptera)." *Journal of the Royal Society Interface* 6 (Suppl 2): S165–S184.

Secor, S. M. 1989. "Adaptive Strategies of Thermoregulation in Freeranging Sidewinders, *Crotalus Cerastes.*"

Seebacher, F., and G. C. Grigg. 1997. "Patterns of Body Temperature in Wild Freshwater Crocodiles, *Crocodylus Johnstoni*: Thermoregulation versus Thermoconformity, Seasonal Acclimatization, and the Effect of Social Interactions." *Copeia*: 549–557.

Setchell, J. M., E. J. Wickings, and L. A. Knapp. 2006. "Signal Content of Red Facial Coloration in Female Mandrills (*Mandrillus Sphinx*)." *Proceedings of the Royal Society B: Biological Sciences* 273 (1599): 2395–2400.

Sherbrooke, W. C., A. J. Scardino, R. De Nys, and L. Schwarzkopf. 2007. "Functional Morphology of Scale Hinges Used to Transport Water: Convergent Drinking Adaptations in Desert Lizards (*Moloch Horridus* and *Phrynosoma Cornutum*)." *Zoomorphology* 126 (2): 89–102.

Shillington, C. 2002. "Thermal Ecology of Male Tarantulas (*Aphonopelma Anax*) During the Mating Season." *Canadian Journal of Zoology* 80 (2) (February): 251–259.

Shuker, K. 2001. *The Hidden Powers of Animals: Uncovering the Secrets of Nature*. London: Marshall Editions.

Sieden, L. S. 2000. *Buckminster Fuller's Universe: His Life and Work*. New York: Basic Books.

Simmaco, M., G. Mignogna, and D. Barra. 1998. "Antimicrobial Peptides from Amphibian Skin: What Do They Tell Us?" *Biopolymers* 47 (6): 435–450.

Smil, V. 2006. *Energy: A Beginner's Guide*. Oxford, UK: Oneworld Publications.

Smith, E. N. 1979. "Behavioral and Physiological Thermoregulation of Crocodilians." *American Zoologist* 19 (1): 239–247.

Smith, N. G. 1983. "Host Plant Toxicity and Migration in the Dayflying Moth *Urania*." *The Florida Entomologist* 66 (1): 76–85.

Spearman, R. I. C. 1967. "On the Nature of the Horny Scales of the Pangolin." *Journal of the Linnean Society of London, Zoology* 46 (310): 267–273.

Stattmann, N. 2003. *Ultra Light — Super Strong: A New Generation of Design Materials*. Basel, Switzerland: Birkhäuser.

Steen, I., and J. B. Steen. 1965. "The Importance of the Legs in the Thermoregulation of Birds." *Acta Physiologica Scandinavica* 63 (3): 285–291.

Stevens, A. 2011. "Introduction to the Basic Drivers of Climate." *Nature Education Knowledge* 2 (2): 6.

Stewart, D. 2004. "The Quest to Quench." *National Wildlife*: 52–63.

Stuart-Fox, D., A. Moussalli, and M. J. Whiting. 2008. "Predator-Specific Camouflage in Chameleons." *Biology Letters* 4 (4): 326–329.

Sunquist, M., and F. Sunquist. 2002. *Wild Cats of the World*. Chicago, IL: University of Chicago Press.

Tattersall, G. J., D. V. Andrade, and A. S. Abe. 2009. "Heat Exchange from the Toucan Bill Reveals a Controllable Vascular Thermal Radiator." *Science* 325 (5939) (July 23): 468–470.

Ternaux, E. 2011. *Material World 3: Innovative Materials for Architecture and Design*. Third ed. Amsterdam: Frame Publishers.

———. 2012. *Industry of Nature: Another Approach to Ecology*. Amsterdam: Frame Publishers.

Terzidis, K. 2006. *Algorithmic Architecture*. Architectural Press.

Thomas, N., and D. Marie. "Fog-Basking Behaviour and Water Collection Efficiency in Namib Desert Darkling Beetles." *Frontiers in Zoology* 7.

Thompson, D. W. 2011. *On Growth and Form*. CreateSpace.

Thompson, J. N. 1994. *The Coevolutionary Process*. Chicago, IL: University of Chicago Press.

———. 2005. *The Geographic Mosaic of Coevolution*. Chicago, IL: University of Chicago Press. 221

Tilder, L., and B. Blostein, Eds. 2009. *Design Ecologies: Sustainable Potentials in Architecture*. New York: Princeton Architectural Press.

Tinbergen, N. 1965. *Animal Behavior*. New York: Time Inc.

Toledo, R. C., and C. Jared. 1993. "Cutaneous Adaptations to Water Balance in Amphibians." *Comparative Biochemistry and Physiology Part A: Physiology* 105 (4): 593–608.

Tomanek, L., and B. Helmuth. 2002. "Physiological Ecology of Rocky Intertidal Organisms: A Synergy of Concepts." *Integrative and Comparative Biology* 42 (4) (August 1): 771–775.

Tong, J., T.-B. Lü, Y.-H. Ma, H.-K. Wang, L.-Q. Ren, and R. D. Arnell. 2007. "Two-Body Abrasive Wear of the Surfaces of Pangolin Scales." *Journal of Bionic Engineering* 4 (2) (June): 77–84.

Tripodi, A. D., and A. L. Szalanski. 2011. "Further Range Extension of *Xylocopa Micans* Lepeletier (Hymenoptera: Apidae)." *Journal of the Kansas Entomological Society* 84 (2): 163–164.

Tucci, Fo. 2009. *Tecnologia e natura. Gli insegnamenti del mondo naturale per il progetto dell'architettura bioclimatica*. Florence, Italy: Alinea Editrice.

Turner, J. S. 2002. *The Extended Organism: The Physiology of Animal-Built Structures*. Cambridge, MA: Harvard University Press.

———. 2010. *The Tinkerer's Accomplice: How Design Emerges from Life Itself*. Cambridge, MA: Harvard University Press.

Tzonis, A., L. Lefaivre, and B. Stagno. 2001. *Tropical Architecture: Critical Regionalism in the Age of Globalization*. Chichester, UK: Academy Press.

U.S. Department of Energy. "EnergyPlus Energy Simulation Software" Last modified March 11, 2011. http://apps1.eere.energy.gov/buildings/ energyplus/cfm/weather_data.cfm

Vattam, S., M. E. Helms, A. Goel, J. Yen, and M. J. Weissburg. In Press. "Learning about and through Biologically Inspired Design." Proceedings Second Design Creativity Workshop Atlanta.

Vincent, J. F. V., O. A. Bogatyreva, N. R. Bogatyrev, A. Bowyer, and A.-K. Pahl. 2006. "Biomimetics: Its Practice and Theory." *Journal of the Royal Society Interface* 3 (9) (August 22): 471–482.

Vincent, J. F.V. 1990. *Structural Biomaterials: Revised*. New York: Princeton University Press.

Vogel, S. 1988. *Life's Devices: The Physical World of Animals and Plants*. New York: Princeton University Press.

———. 2000. Cats' Paws and Catapults: Mechanical Worlds of Nature and People. New York: W. W. Norton & Company.

Vulinec, K. 1997. "Iridescent Dung Beetles: A Different Angle." *Florida Entomologist*: 132–141.

Waldschmidt, S. 1980. "Orientation to the Sun by the Iguanid Lizards *Uta Stansburiana* and *Sceloporus Undulatus*: Hourly and Monthly Variations." *Copeia*: 458–462. 222

Wang, L. C., D. L. Jones, R. A. MacArthur, and W. A. Fuller. 1973. "Adaptation to Cold: Energy Metabolism in an Atypical Lagomorph, the Arctic Hare (*Lepus Arcticus*)." *Canadian Journal of Zoology* 51 (8) (August): 841–846.

Warriner, M. D. 2010. "A Range Extension for the Large Carpenter Bee *Xylocopa Micans* (Hymenoptera: Apidae) with Notes on Floral and Habitat Associations." *Journal of the Kansas Entomological Society* 83 (3): 267–269.

Weissenböck, N., C. M. Weiss, H. M. Schwammer, and H. Kratochvil. 2010. "Thermal Windows on the Body Surface of African Elephants (*Loxodonta Africana*) Studied by Infrared Thermography." *Journal of Thermal Biology* 35 (4) (May): 182–188.

Welch, J. J. 2010. "The 'Island Rule' and Deep-Sea Gastropods: Re-Examining the Evidence." *PloS One* 5 (1): e8776.

Williams, T. M. 1990. "Heat Transfer in Elephants: Thermal Partitioning Based on Skin Temperature Profiles." *Journal of Zoology* 222 (2): 235–245.

Wilson, D. E., and D. M. Reeder, Eds. 2005. *Mammal Species of the World: A Taxonomic and Geographic Reference*, 2-volume Set. 3rd ed. Baltimore, MD: The Johns Hopkins University Press.

Wilson, E. O. 1984. *Biophilia*. Cambridge, MA: Harvard University Press.

Wiscombe, T. 2009. *Emergent:Structural Ecologies*. Wuhan, China: Aadcu Hust Press.

———. 2012. "Beyond Assemblies: System Convergence and Multimateriality." *Bioinspiration & Biomimetics* 7 (1) (March 1): 015001.

Wollenberg, K. C., S. Lötters, C. Mora-Ferrer, and M. Veith. 2008. "Disentangling Composite Colour Patterns in a Poison Frog Species." *Biological Journal of the Linnean Society* 93 (3): 433–444.

Wollenberg, K. C., M. Veith, B. P. Noonan, S. Lötters, and J. M. Quattro. 2006. "Polymorphism versus Species Richness-Systematics of Large Dendrobates from the Eastern Guiana Shield (Amphibia: Dendrobatidae)." *Copeia* 2006 (4): 623–629.

Yoshioka, S., and S. Kinoshita. 2007. "Polarization-Sensitive Color Mixing in the Wing of the Madagascan Sunset Moth." *Optics Express* 15 (5): 2691–2701.

Yoshioka, S., T. Nakano, Y. Nozue, and S. Kinoshita. 2008. "Coloration Using Higher Order Optical Interference in the Wing Pattern of the Madagascan Sunset Moth." *Journal of the Royal Society Interface* 5 (21): 457–464.

Zaera-Polo, A. 2008. *The Politics of the Envelope: A Political Critique of Materialism*. Volume 17.

Author Biographies

(Photograph courtesy of E. Oprandi & S. Yanoviak, respectively)

Ilaria Mazzoleni is an architect and the founder of IM Studio Milano/Los Angeles. Her conceptual work has been published globally, and her built work can be found in Italy, California, and Ghana. Ilaria has gained attention in the fields of sustainable architecture and biomimicry. This has led to her being invited to participate in multiple international conferences and workshops and her written contributions are published in several international architectural magazines. Since 2005 she has been a full-time faculty member at the Southern California Institute of Architecture (SCI-Arc) in Los Angeles. Her professional and academic investigation relates to sustainable architecture on all scales of design with a research focus on biomimicry, where innovation in architecture and design is inspired by the processes and functions of nature. Collaborating with biologists and other scientists from top research institutions, her projects explore the connections between biotic and abiotic elements within ecosystems in order to develop sustainable urban planning strategies and address solutions to global climate change. An ongoing research program has centered on understanding how organisms have evolved and adapted to their environment, and applying that knowledge to design building façades. Current investigations use design as a vehicle to promote awareness about endangered species and emphasize the importance of biodiversity in regions around the world. Mazzoleni explores the performative capacities of organic systems using an analytical approach and the strategy of juxtaposition of real and digital space. The conceptual implications arising from biomimetics and design have led to a body of work that investigates innovative material processes, forms, geometries and structural patterns.

Shauna Price is an evolutionary biologist focusing on speciation in neotropical insects. Her research examines the historical and ecological factors contributing to the high species diversity found in ants with the use of genetic tools, geological data, and morphological analyses. In addition to conducting research, Shauna has collaborated with Ilaria Mazzoleni and IM Studio MI/LA on multiple bio-inspired design projects. She contributes a strong background in ecology and evolution to these studies, with the perspective that inspiration in architecture and design can stem from organisms as small as microbes to broad, ecosystem scales. In particular, symbiotic relationships—close, ongoing associations that have co-evolved between different species—inform her perspective in merging biology with design.

Index

234

Printed and bound by CPI Group (UK) Ltd, Croydon, CR0 4YY

18/10/2024

01776253-0001